D0849006

Nadia Nedjah, Ajith Abraham, Luiza de Macedo Mourelle (Eds.)

Genetic Systems Programming

Studies in Computational Intelligence, Volume 13

Editor-in-chief
Prof. Janusz Kacprzyk
Systems Research Institute
Polish Academy of Sciences
ul. Newelska 6
01-447 Warsaw
Poland
E-mail: kacprzyk@ibspan.waw.pl

Further volumes of this series
can be found on our homepage:
springer.com

Vol. 1. Tetsuya Hoya
Artificial Mind System – Kernel Memory Approach, 2005
ISBN 3-540-26072-2

Vol. 2. Saman K. Halgamuge, Lipo Wang (Eds.)
Computational Intelligence for Modelling and Prediction, 2005
ISBN 3-540-26071-4

Vol. 3. Bożena Kostek
Perception-Based Data Processing in Acoustics, 2005
ISBN 3-540-25729-2

Vol. 4. Saman K. Halgamuge, Lipo Wang (Eds.)
Classification and Clustering for Knowledge Discovery, 2005
ISBN 3-540-26073-0

Vol. 5. Da Ruan, Guoqing Chen, Etienne E. Kerre, Geert Wets (Eds.)
Intelligent Data Mining, 2005
ISBN 3-540-26256-3

Vol. 6. Tsau Young Lin, Setsuo Ohsuga, Churn-Jung Liau, Xiaohua Hu, Shusaku Tsumoto (Eds.)
Foundations of Data Mining and Knowledge Discovery, 2005
ISBN 3-540-26257-1

Vol. 7. Bruno Apolloni, Ashish Ghosh, Ferda Alpaslan, Lakhmi C. Jain, Srikanta Patnaik (Eds.)
Machine Learning and Robot Perception, 2005
ISBN 3-540-26549-X

Vol. 8. Srikanta Patnaik, Lakhmi C. Jain, Spyros G. Tzafestas, Germano Resconi, Amit Konar (Eds.)
Innovations in Robot Mobility and Control, 2005
ISBN 3-540-26892-8

Vol. 9. Tsau Young Lin, Setsuo Ohsuga, Churn-Jung Liau, Xiaohua Hu (Eds.)
Foundations and Novel Approaches in Data Mining, 2005
ISBN 3-540-28315-3

Vol. 10. Andrzej P. Wierzbicki, Yoshiteru Nakamori
Creative Space, 2005
ISBN 3-540-28458-3

Vol. 11. Antoni Ligêza
Logical Foundations for Rule-Based Systems, 2006
ISBN 3-540-29117-2

Vol. 13. Nadia Nedjah, Ajith Abraham, Luiza de Macedo Mourelle (Eds.)
Genetic Systems Programming, 2006
ISBN 3-540-29849-5

Nadia Nedjah
Ajith Abraham
Luiza de Macedo Mourelle
(Eds.)

Genetic Systems Programming

Theory and Experiences

Dr. Nadia Nedjah
Dr. Luiza de Macedo Mourelle
Faculdade de Engenharia
Universidade do Estado
do Rio de Janeiro
Rua São Francisco Xavier
524, 20550-900 Maracanã, Rio de Janeiro
Brazil
E-mail: nadia@eng.uerj.br

Dr. Ajith Abraham
School for Computer Science
and, Engineering
Chung-Ang University
Heukseok-dong 221
156-756 Seoul, Korea
Republic of (South Korea)
E-mail: ajith.abraham@ieee.org

Library of Congress Control Number: 2005936350

ISSN print edition: 1860-949X
ISSN electronic edition: 1860-9503
ISBN-10 3-540-29849-5 Springer Berlin Heidelberg New York
ISBN-13 978-3-540-29849-6 Springer Berlin Heidelberg New York

Springer is a part of Springer Science+Business Media
springer.com

Printed in The Netherlands

Typesetting: by the authors and TechBooks using a Springer LaTeX macro package

Printed on acid-free paper SPIN: 11521433 89/TechBooks 5 4 3 2 1 0

Foreword

The editors of this volume, Nadia Nedjah, Ajith Abraham and Luiza de Macedo Mourelle, have done a superb job of assembling some of the most innovative and intriguing applications and additions to the methodology and theory of genetic programming – an automatic programming technique that starts from a high-level statement of what needs to be done and automatically creates a computer program to solve the problem.

When the genetic algorithm first appeared in the 1960s and 1970s, it was an academic curiosity that was primarily useful in understanding certain aspects of how evolution worked in nature. In the 1980s, in tandem with the increased availability of computing power, practical applications of genetic and evolutionary computation first began to appear in specialized fields. In the 1990s, the relentless iteration of Moore's law – which tracks the 100-fold increase in computational power every 10 years – enabled genetic and evolutionary computation to deliver the first results that were comparable and competitive with the work of creative humans. As can be seen from the preface and table of contents, the field has already begun the 21st century with a cornucopia of applications, as well as additions to the methodology and theory, including applications to information security systems, compilers, data mining systems, stock market prediction systems, robotics, and automatic programming.

Looking forward three decades, there will be a 1,000,000-fold increase in computational power. Given the impressive human-competitive results already delivered by genetic programming and other techniques of evolutionary computation, the best is yet to come.

September 2005 *Professor John R. Koza*

Preface

Designing complex programs such as operating systems, compilers, filing systems, data base systems, etc. is an old ever lasting research area. Genetic programming is a relatively new promising and growing research area. Among other uses, it provides efficient tools to deal with hard problems by evolving creative and competitive solutions. Systems Programming is generally strewn with such hard problems. This book is devoted to reporting innovative and significant progress about the contribution of genetic programming in systems programming. The contributions of this book clearly demonstrate that genetic programming is very effective in solving hard and yet-open problems in systems programming. Followed by an introductory chapter, in the remaining contributed chapters, the reader can easily learn about systems where genetic programming can be applied successfully. These include but are not limited to, information security systems (see Chapter 3), compilers (see Chapter 4), data mining systems (see Chapter 5), stock market prediction systems (see Chapter 6), robots (see Chapter 8) and automatic programming (see Chapters 7 and 9).

In Chapter 1, which is entitled *Evolutionary Computation: from Genetic Algorithms to Genetic Programming*, the authors introduce and review the development of the field of evolutionary computations from standard genetic algorithms to genetic programming, passing by evolution strategies and evolutionary programming. The main differences among the different evolutionary computation techniques are also illustrated in this Chapter.

In Chapter 2, which is entitled *Automatically Defined Functions in Gene Expression Programming*, the author introduces the cellular system of Gene Expression Programming with Automatically Defined Functions (ADF) and discusses the importance of ADFs in Automatic Programming by comparing the performance of sophisticated learning systems with ADFs with much simpler ones without ADFs on a benchmark problem of symbolic regression.

In Chapter 3, which is entitled *Evolving Intrusion Detection Systems*, the authors present an Intrusion Detection System (IDS), which is a program that analyzes what happens or has happened during an execution and tries to find indications that the computer has been misused. An IDS does not eliminate the use of preventive mechanism but it works as the last defensive mechanism in securing the system. The authors also evaluate the performances of two Genetic Programming techniques for IDS namely Linear Genetic Programming (LGP) and Multi-Expression Programming (MEP). They compare the obtained results with some machine learning techniques like Support Vector Machines (SVM) and Decision Trees (DT). The authors claim that empirical results clearly show that GP techniques could play an important role in designing real time intrusion detection systems.

In Chapter 4, which is entitled *Evolutionary Pattern Matching Using Genetic Programming*, the authors apply GP to the hard problem of engineering pattern matching automata for non-sequential pattern set, which is almost always the case in functional programming. They engineer good traversal orders that allow one to design an efficient adaptive pattern-matchers that visit necessary positions only. The authors claim that doing so the evolved pattern matching automata improves time and space requirements of pattern-matching as well as the termination properties of term evaluation.

In Chapter 5, which is entitled *Genetic Programming in Data Modelling*, the author demonstrates some abilities of Genetic Programming (GP) in Data Modelling (DM). The author shows that GP can make data collected in large databases more useful and understandable. The author concentrates on mathematical modelling, classification, prediction and modelling of time series.

In Chapter 6, which is entitled *Stock Market Modeling Using Genetic Programming Ensembles*, the authors introduce and use two Genetic Programming (GP) techniques: Multi-Expression Programming (MEP) and Linear Genetic Programming (LGP) for the prediction of two stock indices. They compare the performance of the GP techniques with an artificial neural network trained using Levenberg-Marquardt algorithm and Takagi-Sugeno neuro-fuzzy model. As a case study, the authors consider Nasdaq-100 index of Nasdaq Stock Market and the S&P CNX NIFTY stock index as test data. Based on the empirical results obtained the authors conclude that Genetic Programming techniques are promising methods for stock prediction. Finally, they formulate an ensemble of these two techniques using a multiobjective evolutionary algorithm and claim that results reached by ensemble of GP techniques are better than the results obtained by each GP technique individually.

In Chapter 7, which is entitled *Evolutionary Digital Circuit Design Using Genetic Programming*, the authors study two different circuit encodings used for digital circuit evolution. The first approach is based on genetic programming, wherein digital circuits consist of their data flow based specifications. In this approach, individuals are internally represented by the abstract trees/DAG of the corresponding circuit specifications. In the second approach, digital circuits are thought of as a map of rooted gates. So individuals are represented by two-dimensional arrays of cells. The authors compare the impact of both individual representations on the evolution process of digital circuits. The authors reach the conclusion that employing either of these approaches yields circuits of almost the same characteristics in terms of space and response time. However, the evolutionary process is much shorter with the second linear encoding.

In Chapter 8, which is entitled *Evolving Complex Robotic Behaviors Using Genetic Programming*, the author reviews different methods for evolving complex robotic behaviors. The methods surveyed use two different approaches: The first one introduces hierarchy into GP by using library of procedures or new primitive functions and the second one uses GP to evolve the building modules of robot controller hierarchy. The author comments on including practical issues of evolution as well as comparison between the two approaches.

In Chapter 9, which is entitled *Automatic Synthesis of Microcontroller Assembly Code Through Linear Genetic Programming*, the authors focus on the potential of linear genetic programming in the automatic synthesis of microcontroller assembly language programs. For them, these programs implement strategies for time-optimal or sub-optimal control of the system to be controlled, based on mathematical modeling through dynamic equations. They also believe that within this application class, the best model is the one used in linear genetic programming, in which each chromosome is represented by an instruction list. The authors find the synthesis of programs that implement optimal-time control strategies for microcontrollers, directly in assembly language, as an attractive alternative that overcomes the difficulties presented by the conventional design of optimal control systems. This chapter widens the perspective of broad usage of genetic programming in automatic control.

We are very much grateful to the authors of this volume and to the reviewers for their tremendous service by critically reviewing the chapters. The editors would like also to thank Prof. Janusz Kacprzyk, the editor-in-chief of the Studies in Computational Intelligence Book Series and Dr. Thomas Ditzinger, Springer Verlag, Germany for the editorial assistance and excellent cooperative collaboration to produce this important scientific work. We hope that the reader will share our excitement to present this volume on **Genetic Systems Programming** and will find it useful.

Brazil
August 2005

Nadia Nedjah
Ajith Abraham
Luiza M. Mourelle

Contents

List of Figures

List of Tables

1

Evolutionary Computation: from Genetic Algorithms to Genetic Programming

Ajith Abraham[1], Nadia Nedjah[2], and Luiza de Macedo Mourelle[3]

[1] School of Computer Science and Engineering Chung-Ang University 410, 2nd Engineering Building 221, Heukseok-dong, Dongjak-gu Seoul 156-756, Korea
ajith.abraham@ieee.org, http://www.ajith.softcomputing.net

[2] Department of Electronics Engineering and Telecommunications, Engineering Faculty, State University of Rio de Janeiro, Rua São Francisco Xavier, 524, Sala 5022-D, Maracanã, Rio de Janeiro, Brazil
nadia@eng.uerj.br, http://www.eng.uerj.br/~nadia

[3] Department of System Engineering and Computation, Engineering Faculty, State University of Rio de Janeiro, Rua São Francisco Xavier, 524, Sala 5022-D, Maracanã, Rio de Janeiro, Brazil
ldmm@eng.uerj.br, http://www.eng.uerj.br/~ldmm

Evolutionary computation, offers practical advantages to the researcher facing difficult optimization problems. These advantages are multi-fold, including the simplicity of the approach, its robust response to changing circumstance, its flexibility, and many other facets. The evolutionary approach can be applied to problems where heuristic solutions are not available or generally lead to unsatisfactory results. As a result, evolutionary computation have received increased interest, particularly with regards to the manner in which they may be applied for practical problem solving.

In this chapter, we review the development of the field of evolutionary computations from standard genetic algorithms to genetic programming, passing by evolution strategies and evolutionary programming. For each of these orientations, we identify the main differences from the others. We also, describe the most popular variants of genetic programming. These include linear genetic programming (LGP), gene expression programming (GEP), multi-expresson programming (MEP), Cartesian genetic programming (CGP), traceless genetic programming (TGP), gramatical evolution (GE) and genetic glgorithm for deriving software (GADS).

A. Abraham et al.: *Evolutionary Computation: from Genetic Algorithms to Genetic Programming*, Studies in Computational Intelligence (SCI) **13**, 1–20 (2006)
www.springerlink.com

1.1 Introduction

In nature, evolution is mostly determined by natural selection or different individuals competing for resources in the environment. Those individuals that are better are more likely to survive and propagate their genetic material. The encoding for genetic information (genome) is done in a way that admits asexual reproduction which results in offspring that are genetically identical to the parent. Sexual reproduction allows some exchange and re-ordering of chromosomes, producing offspring that contain a combination of information from each parent. This is the recombination operation, which is often referred to as crossover because of the way strands of chromosomes cross over during the exchange. The diversity in the population is achieved by mutation.

Evolutionary algorithms are ubiquitous nowadays, having been successfully applied to numerous problems from different domains, including optimization, automatic programming, machine learning, operations research, bioinformatics, and social systems. In many cases the mathematical function, which describes the problem is not known and the values at certain parameters are obtained from simulations. In contrast to many other optimization techniques an important advantage of evolutionary algorithms is they can cope with multi-modal functions.

Usually grouped under the term evolutionary computation [1] or evolutionary algorithms, we find the domains of genetic algorithms [9], evolution strategies [17, 19], evolutionary programming [5] and genetic programming [11]. They all share a common conceptual base of simulating the evolution of individual structures via processes of selection, mutation, and reproduction. The processes depend on the perceived performance of the individual structures as defined by the problem.

A population of candidate solutions (for the optimization task to be solved) is initialized. New solutions are created by applying reproduction operators (mutation and/or crossover). The fitness (how good the solutions are) of the resulting solutions are evaluated and suitable selection strategy is then applied to determine which solutions will be maintained into the next generation. The procedure is then iterated and is illustrated in Fig. 1.1.

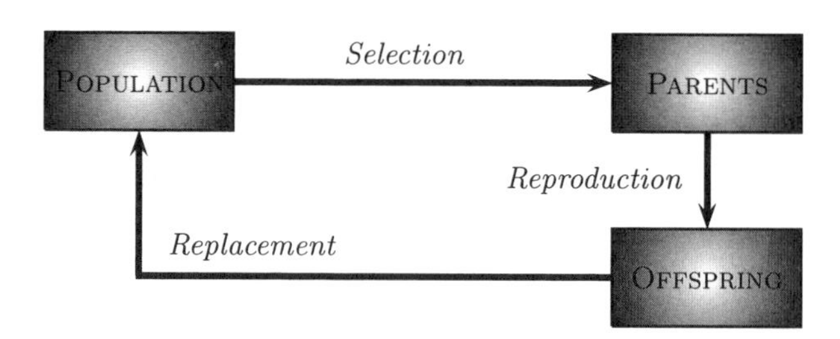

Fig. 1.1. Flow chart of an evolutionary algorithm

1.1.1 Advantages of Evolutionary Algorithms

A primary advantage of evolutionary computation is that it is conceptually simple. The procedure may be written as difference equation (1.1):

$$x[t+1] = s(v(x[t])) \tag{1.1}$$

where $x[t]$ is the population at time t under a representation x, v is a random variation operator, and s is the selection operator [6].

Other advantages can be listed as follows:

- Evolutionary algorithm performance is representation independent, in contrast with other numerical techniques, which might be applicable for only continuous values or other constrained sets.
- Evolutionary algorithms offer a framework such that it is comparably easy to incorporate prior knowledge about the problem. Incorporating such information focuses the evolutionary search, yielding a more efficient exploration of the state space of possible solutions.
- Evolutionary algorithms can also be combined with more traditional optimization techniques. This may be as simple as the use of a gradient minimization used after primary search with an evolutionary algorithm (for example fine tuning of weights of a evolutionary neural network), or it may involve simultaneous application of other algorithms (e.g., hybridizing with simulated annealing or tabu search to improve the efficiency of basic evolutionary search).
- The evaluation of each solution can be handled in parallel and only selection (which requires at least pair wise competition) requires some serial processing. Implicit parallelism is not possible in many global optimization algorithms like simulated annealing and Tabu search.
- Traditional methods of optimization are not robust to dynamic changes in problem the environment and often require a complete restart in order to provide a solution (e.g., dynamic programming). In contrast, evolutionary algorithms can be used to adapt solutions to changing circumstance.
- Perhaps the greatest advantage of evolutionary algorithms comes from the ability to address problems for which there are no human experts. Although human expertise should be used when it is available, it often proves less than adequate for automating problem-solving routines.

1.2 Genetic Algorithms

A typical flowchart of a Genetic Algorithm (GA) is depicted in Fig. 1.2. One iteration of the algorithm is referred to as a generation. The basic GA is very generic and there are many aspects that can be implemented differently according to the problem (For instance, representation of solution or chromosomes, type of encoding, selection strategy, type of crossover and mutation operators, etc.) In practice, GAs are implemented by having arrays of bits or characters to represent the chromosomes. The individuals in the population then go through a process of simulated evolution. Simple bit manipulation operations allow the implementation of crossover, mutation and other operations. The number of bits for every gene (parameter) and the decimal range in which they decode are usually the same but nothing precludes the utilization of a different number of bits or range for every gene.

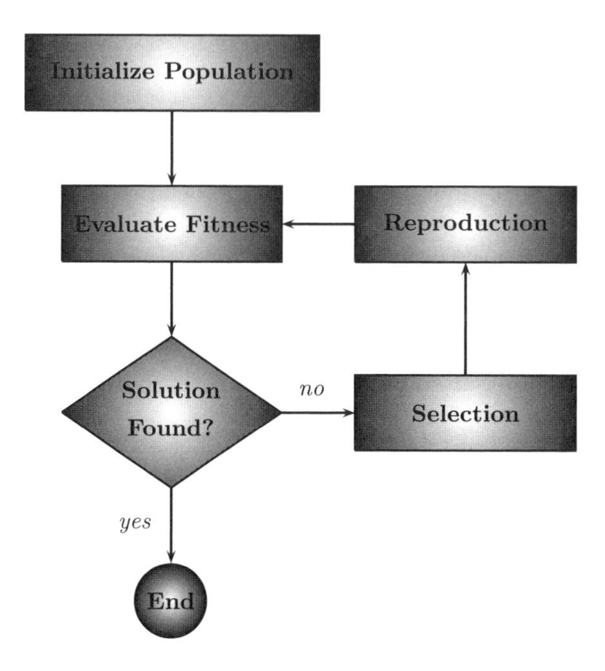

Fig. 1.2. Flow chart of basic genetic algorithm iteration

When compared to other evolutionary algorithms, one of the most important GA feature is its focus on fixed-length character strings although variable-length strings and other structures have been used.

1.2.1 Encoding and Decoding

In a typical application of GA's, the given problem is transformed into a set of genetic characteristics (parameters to be optimized) that will survive in the best possible manner in the environment. Example, if the task is to optimize the function given in 1.2.

$$\min f(x_1, x_2) = (x_1 - 5)^2 + (x_2 - 2)^2, -3 \leq x_1 \leq 3,\ -8 \leq x_2 \leq 8 \quad (1.2)$$

The parameters of the search are identified as x_1 and x_2, which are called the phenotypes in evolutionary algorithms. In genetic algorithms, the phenotypes (parameters) are usually converted to genotypes by using a coding procedure. Knowing the ranges of x_1 and x_2 each variable is to be represented using a suitable binary string. This representation using binary coding makes the parametric space independent of the type of variables used. The genotype (chromosome) should in some way contain information about solution, which is also known as encoding. GA's use a binary string encoding as shown below.

Chromosome A: 110110111110100110110
Chromosome B: 110111101010100011110

Each bit in the chromosome strings can represent some characteristic of the solution. There are several types of encoding (example, direct integer or real numbers encoding). The encoding depends directly on the problem.

Permutation encoding can be used in ordering problems, such as Travelling Salesman Problem (TSP) or task ordering problem. In permutation encoding, every chromosome is a string of numbers, which represents number in a sequence. A chromosome using permutation encoding for a 9 city TSP problem will look like as follows:

Chromosome A: 4 5 3 2 6 1 7 8 9
Chromosome B: 8 5 6 7 2 3 1 4 9

Chromosome represents order of cities, in which salesman will visit them. Special care is to taken to ensure that the strings represent real sequences after crossover and mutation. Floating-point representation is very useful for numeric optimization (example: for encoding the weights of a neural network). It should be noted that in many recent applications more sophisticated genotypes are appearing (example: chromosome can be a tree of symbols, or is a combination of a string and a tree, some parts of the chromosome are not allowed to evolve etc.)

1.2.2 Schema Theorem and Selection Strategies

Theoretical foundations of evolutionary algorithms can be partially explained by schema theorem [9], which relies on the concept of schemata. Schemata are templates that partially specify a solution (more strictly, a solution in the genotype space). If genotypes are strings built using symbols from an alphabet A, schemata are strings whose symbols belong to $A \cup \{*\}$. This extra-symbol * must be interpreted as a *wildcard*, being loci occupied by it called undefined. A chromosome is said to match a schema if they agree in the defined positions.

For example, the string 10011010 matches the schemata 1******* and **011*** among others, but does not match *1*11*** because they differ in the second gene (the first defined gene in the schema). A schema can be viewed as a hyper-plane in a k-dimensional space representing a set of solutions with common properties. Obviously, the number of solutions that match a schema H depend on the number of defined positions in it. Another related concept is the *defining-length* of a schema, defined as the distance between the first and the last defined positions in it. The GA works by allocating strings to best schemata exponentially through successive generations, being the selection mechanism the main responsible for this behaviour. On the other hand the crossover operator is responsible for exploring new combinations of the present schemata in order to get the fittest individuals. Finally the purpose of the mutation operator is to introduce fresh genotypic material in the population.

1.2.3 Reproduction Operators

Individuals for producing offspring are chosen using a selection strategy after evaluating the fitness value of each individual in the selection pool. Each individual in the selection pool receives a reproduction probability depending on its own fitness value and the fitness value of all other individuals in the selection pool. This fitness is used for the actual selection step afterwards. Some of the popular selection schemes are discussed below.

Roulette Wheel Selection

The simplest selection scheme is roulette-wheel selection, also called stochastic sampling with replacement. This technique is analogous to a roulette wheel with each slice proportional in size to the fitness. The individuals are mapped to contiguous segments of a line, such that each individual's segment is equal in size to its fitness. A random number is generated and the individual whose segment spans the random number is selected. The process is repeated until the desired number of individuals is obtained. As illustrated in Fig. 1.3, chromosome_1 has the highest probability for being selected since it has the highest fitness.

Tournament Selection

In tournament selection a number of individuals is chosen randomly from the population and the best individual from this group is selected as parent. This process is repeated as often as individuals to choose. These selected parents produce uniform at random offspring. The tournament size will often depend on the problem, population size etc. The parameter for tournament selection is the tournament size. Tournament size takes values ranging from 2 – number of individuals in population.

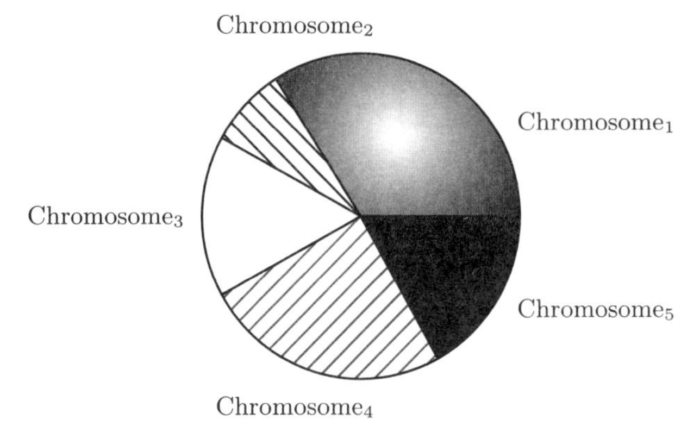

Fig. 1.3. Roulette wheel selection

Elitism

When creating new population by crossover and mutation, we have a big chance that we will lose the best chromosome. Elitism is name of the method that first copies the best chromosome (or a few best chromosomes) to new population. The rest is done in classical way. Elitism can very rapidly increase performance of GA, because it prevents losing the best-found solution.

Genetic Operators

Crossover and mutation are two basic operators of GA. Performance of GA very much depends on the genetic operators. Type and implementation of operators depends on encoding and also on the problem. There are many ways how to do crossover and mutation. In this section we will demonstrate some of the popular methods with some examples and suggestions how to do it for different encoding schemes.

Crossover. It selects genes from parent chromosomes and creates a new offspring. The simplest way to do this is to choose randomly some crossover point and everything before this point is copied from the first parent and then everything after a crossover point is copied from the second parent. A single point crossover is illustrated as follows (| is the crossover point):

Chromosome A:	**11111** | **00100110110**
Chromosome B:	10011 | 11000011110
Offspring A:	**11111** | 11000011110
Offspring B:	10011 | **00100110110**

As illustrated in Fig. 1.4, there are several crossover techniques. In a uniform crossover bits are randomly copied from the first or from the second

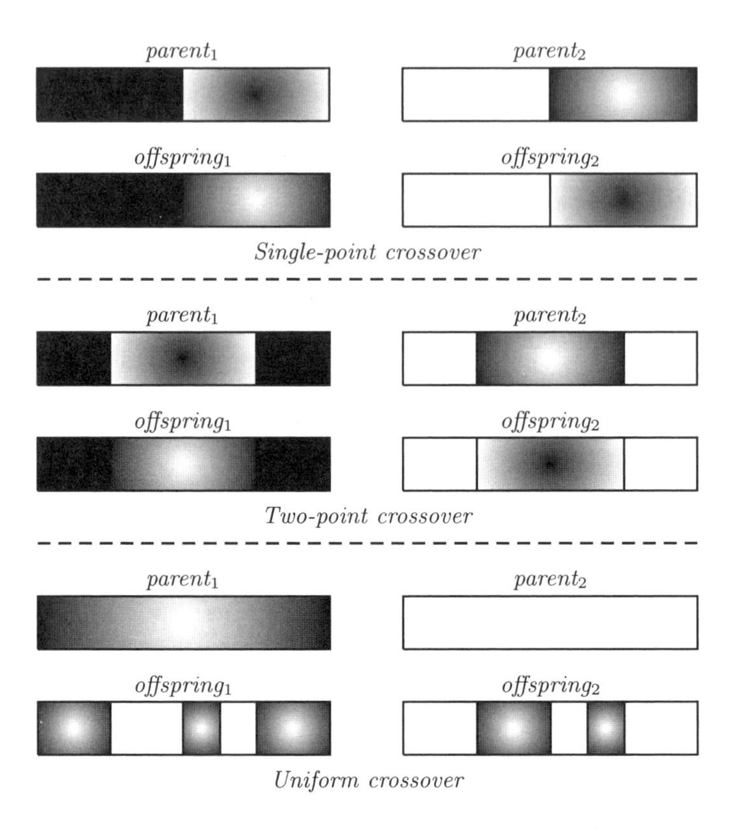

Fig. 1.4. Types of crossover operators

parent. Specific crossover made for a specific problem can improve the GA performance.

Mutation. After crossover operation, mutation takes place. Mutation changes randomly the new offspring. For binary encoding mutation is performed by changing a few randomly chosen bits from 1 to 0 or from 0 to 1. Mutation depends on the encoding as well as the crossover. For example when we are encoding permutations, mutation could be exchanging two genes. A simple mutation operation is illustrated as follows:

Chromosome A: 110**1**111000011110
Chromosome B: 110110**0**100110110

Offspring A: 110**0**111000011110
Offspring B: 1101101**1**00110110

For many optimization problems there may be multiple, equal, or unequal optimal solutions. Sometimes a simple GA cannot maintain stable populations at different optima of such functions. In the case of unequal optimal solutions, the population invariably converges to the global optimum. *Niching* helps to maintain subpopulations near global and local optima. A niche is viewed as an organism's environment and a species as a collection of organisms with similar features. Niching helps to maintain subpopulations near global and local optima by introducing a controlled competition among different solutions near every local optimal region. Niching is achieved by a sharing function, which creates subdivisions of the environment by degrading an organism's fitness proportional to the number of other members in its neighbourhood. The amount of sharing contributed by each individual into its neighbour is determined by their proximity in the decoded parameter space (phenotypic sharing) based on a distance measure.

1.3 Evolution Strategies

Evolution Strategy (ES) was developed by Rechenberg [17] at Technical University, Berlin. ES tend to be used for empirical experiments that are difficult to model mathematically. The system to be optimized is actually constructed and ES is used to find the optimal parameter settings. Evolution strategies merely concentrate on translating the fundamental mechanisms of biological evolution for technical optimization problems. The parameters to be optimized are often represented by a vector of real numbers (object parameters – op). Another vector of real numbers defines the strategy parameters (sp) which controls the mutation of the objective parameters. Both object and strategic parameters form the data-structure for a single individual. A population P of n individuals could be described as $P = (c_1, c_2, \ldots, c_{n-1}, c_n)$, where the ith chromosome c_i is defined as $c_i = (op, sp)$ with $op = (o_1, o_2, ..., o_{n-1}, o_n)$ and $sp = (s_1, s_2, ..., s_{n-1}, s_n)$.

1.3.1 Mutation in Evolution Strategies

The mutation operator is defined as component wise addition of normal distributed random numbers. Both the objective parameters and the strategy parameters of the chromosome are mutated. A mutant's object-parameters vector is calculated as $o_p(mut) = o_p + N_0(s_p)$, where $N_0(s_i)$ is the Gaussian distribution of mean-value 0 and standard deviation s_i. Usually the strategy parameters mutation step size is done by adapting the standard deviation s_i. For instance, this may be done by $s_p(mut) = (s_1 \times A_1, s_2 \times A_2, \ldots, s_{n-1} \times A_{n-1}, s_n \times A_n)$, where A_i is randomly chosen from α or $1/\alpha$ depending on the value of equally distributed random variable E of [0,1] with $A_i = \alpha$ if $E < 0.5$ and $A_i = 1/\alpha$ if $E \geq 0.5$. The parameter α is usually referred to as *strategy parameters adaptation value.*

1.3.2 Crossover (Recombination) in Evolution Strategies

For two chromosomes $c_1 = (o_p(c_1), s_p(c_1))$ and $c_2 = (o_p(c_2), s_p(c_2))$ the crossover operator is defined $R(c_1, c_2) = c = (o_p, s_p)$ with $o_p(i) = (o_p(c_1), i|o_p(c_2), i)$ and $s_p(i) = (s_p(c_1), i|s_p(c_2), i)$. By defining $o_p(i)$ and $s_p(i) = (x|y)$ a value is randomly assigned for either x or y (50% selection probability for x and y).

1.3.3 Controling the Evolution

Let P be the number of parents in generation 1 and let C be the number of children in generation i. There are basically four different types of evolution strategies: P, C, $P{+}C$, $P/R, C$ and $P/R{+}C$ as discussed below. They mainly differ in how the parents for the next generation are selected and the usage of crossover operators.

P, C Strategy

The P parents produce C children using mutation. Fitness values are calculated for each of the C children and the best P children become next generation parents. The best individuals of C children are sorted by their fitness value and the first P individuals are selected to be next generation parents ($C \geq P$).

$P + C$ Strategy

The P parents produce C children using mutation. Fitness values are calculated for each of the C children and the best P individuals of both parents and children become next generation parents. Children and parents are sorted by their fitness value and the first P individuals are selected to be next generation parents.

$P/R, C$ Strategy

The P parents produce C children using mutation and crossover. Fitness values are calculated for each of the C children and the best P children become next generation parents. The best individuals of C children are sorted by their fitness value and the first P individuals are selected to be next generation parents ($C \geq P$). Except the usage of crossover operator this is exactly the same as P, C strategy.

$P/R + C$ Strategy

The P parents produce C children using mutation and recombination. Fitness values are calculated for each of the C children and the best P individuals of both parents and children become next generation parents. Children and parents are sorted by their fitness value and the first P individuals are selected to be next generation parents. Except the usage of crossover operator this is exactly the same as $P + C$ strategy.

1.4 Evolutionary Programming

Fogel, Owens and Walsh's book [5] is the landmark publication for Evolutionary Programming (EP). In the book, Finite state automata are evolved to predict symbol strings generated from Markov processes and non-stationary time series. The basic evolutionary programming method involves the following steps:

1. Choose an initial population (possible solutions at random). The number of solutions in a population is highly relevant to the speed of optimization, but no definite answers are available as to how many solutions are appropriate (other than > 1) and how many solutions are just wasteful.
2. New offspring's are created by mutation. Each offspring solution is assessed by computing its fitness. Typically, a stochastic tournament is held to determine N solutions to be retained for the population of solutions. It should be noted that evolutionary programming method typically does not use any crossover as a genetic operator.

When comparing evolutionary programming to genetic algorithm, one can identify the following differences:

1. GA is implemented by having arrays of bits or characters to represent the chromosomes. In EP there are no such restrictions for the representation. In most cases the representation follows from the problem.
2. EP typically uses an adaptive mutation operator in which the severity of mutations is often reduced as the global optimum is approached while GA's use a pre-fixed mutation operator. Among the schemes to adapt the mutation step size, the most widely studied being the "meta-evolutionary" technique in which the variance of the mutation distribution is subject to mutation by a fixed variance mutation operator that evolves along with the solution.

On the other hand, when comparing evolutionary programming to evolution strategies, one can identify the following differences:

1. When implemented to solve real-valued function optimization problems, both typically operate on the real values themselves and use adaptive reproduction operators.

2. EP typically uses stochastic tournament selection while ES typically uses deterministic selection.
3. EP does not use crossover operators while ES (P/R,C and P/R+C strategies) uses crossover. However the effectiveness of the crossover operators depends on the problem at hand.

1.5 Genetic Programming

Genetic Programming (GP) technique provides a framework for automatically creating a working computer program from a high-level problem statement of the problem [11]. Genetic programming achieves this goal of automatic programming by genetically breeding a population of computer programs using the principles of Darwinian natural selection and biologically inspired operations. The operations include most of the techniques discussed in the previous sections. The main difference between genetic programming and genetic algorithms is the representation of the solution. Genetic programming creates computer programs in the LISP or scheme computer languages as the solution. LISP is an acronym for LISt Processor and was developed by John McCarthy in the late 1950s [8]. Unlike most languages, LISP is usually used as an interpreted language. This means that, unlike compiled languages, an interpreter can process and respond directly to programs written in LISP. The main reason for choosing LISP to implement GP is due to the advantage of having the programs and data have the same structure, which could provide easy means for manipulation and evaluation.

Genetic programming is the extension of evolutionary learning into the space of computer programs. In GP the individual population members are not fixed length character strings that encode possible solutions to the problem at hand, they are programs that, when executed, are the candidate solutions to the problem. These programs are expressed in genetic programming as parse trees, rather than as lines of code. For example, the simple program "$a + b * f(4, a, c)$" would be represented as shown in Fig. 1.5. The terminal and function sets are also important components of genetic programming. The terminal and function sets are the alphabet of the programs to be made. The terminal set consists of the variables (example, a,b and c in Fig. 1.5) and constants (example, 4 in Fig. 1.5).

The most common way of writing down a function with two arguments is the infix notation. That is, the two arguments are connected with the operation symbol between them as $a + b$ or $a * b$. A different method is the prefix notation. Here the operation symbol is written down first, followed by its required arguments as $+ab$ or $*ab$. While this may be a bit more difficult or just unusual for human eyes, it opens some advantages for computational uses. The computer language LISP uses symbolic expressions (or S-expressions) composed in prefix notation. Then a simple S-expression could be $(operator, argument)$ where *operator* is the name of a function and

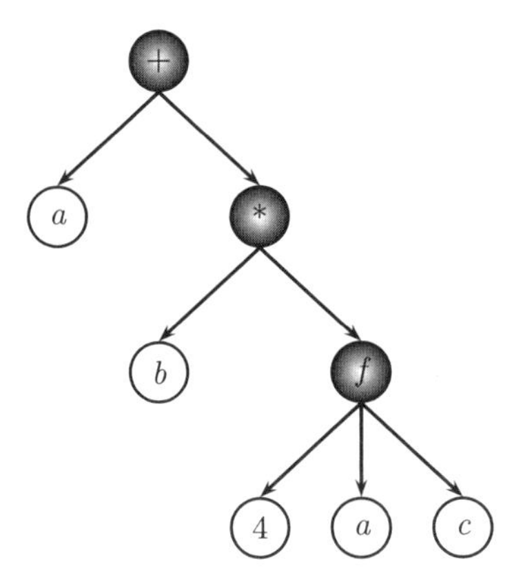

Fig. 1.5. A simple tree structure of GP

argument can be either a constant or a variable or either another symbolic expression as $(operator, argument(operator, argument)(operator, argument))$. Generally speaking, GP procedure could be summarized as follows:

- Generate an initial population of random compositions of the functions and terminals of the problem;
- Compute the fitness values of each individual in the population ;
- Using some selection strategy and suitable reproduction operators produce two offspring;
- Procedure is iterated until the required solution is found or the termination conditions have reached (specified number of generations).

1.5.1 Computer Program Encoding

A parse tree is a structure that grasps the interpretation of a computer program. Functions are written down as nodes, their arguments as leaves. A subtree is the part of a tree that is under an inner node of this tree. If this tree is cut out from its parent, the inner node becomes a root node and the subtree is a valid tree of its own.

There is a close relationship between these parse trees and S-expression; in fact these trees are just another way of writing down expressions. While functions will be the nodes of the trees (or the operators in the S-expressions) and can have other functions as their arguments, the leaves will be formed by terminals, that is symbols that may not be further expanded. Terminals can be variables, constants or specific actions that are to be performed. The process of selecting the functions and terminals that are needed or useful for finding a solution to a given problem is one of the key steps in GP. Evaluation of these structures is straightforward. Beginning at the root node, the values of all sub-expressions (or subtrees) are computed, descending the tree down to the leaves.

1.5.2 Reproduction of Computer Programs

The creation of an offspring from the crossover operation is accomplished by deleting the crossover fragment of the first parent and then inserting the crossover fragment of the second parent. The second offspring is produced in a symmetric manner. A simple crossover operation is illustrated in Fig. 1.6. In GP the crossover operation is implemented by taking randomly selected sub trees in the individuals and exchanging them.

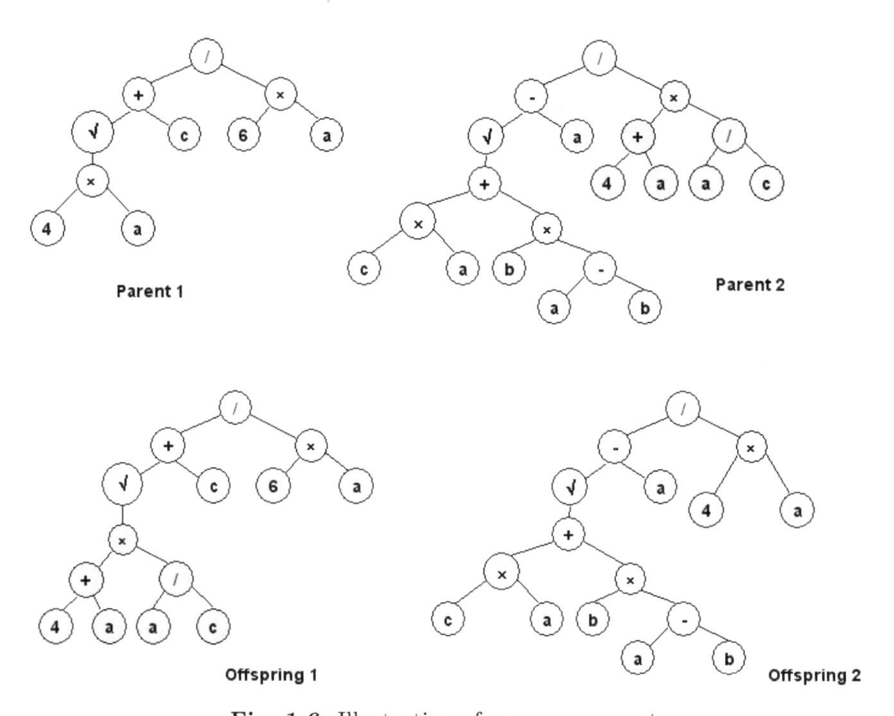

Fig. 1.6. Illustration of crossover operator

Mutation is another important feature of genetic programming. Two types of mutations are commonly used. The simplest type is to replace a function or a terminal by a function or a terminal respectively. In the second kind an entire subtree can replace another subtree. Fig. 1.7 explains the concept of mutation.

GP requires data structures that are easy to handle and evaluate and robust to structural manipulations. These are among the reasons why the class

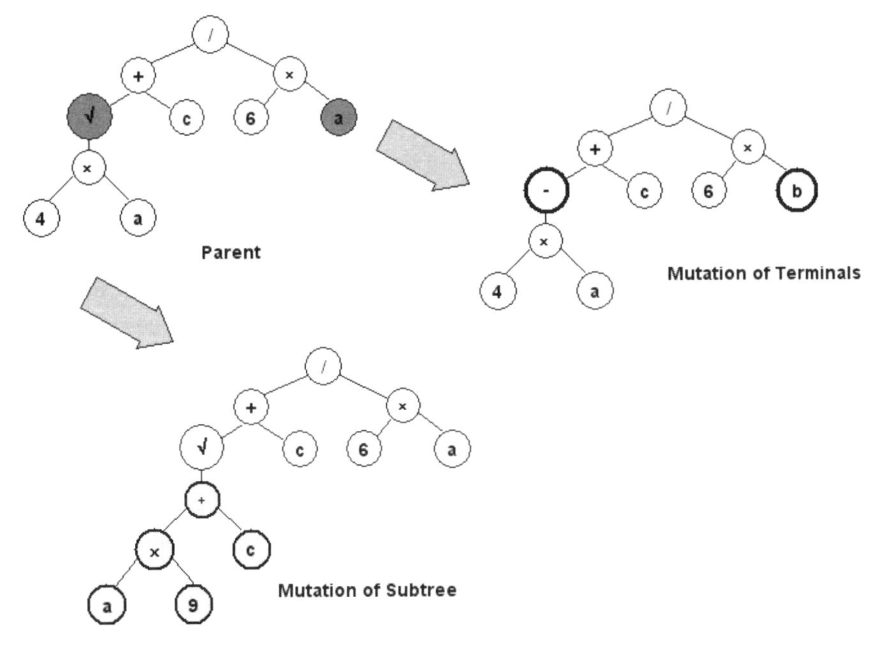

Fig. 1.7. Illustration of mutation operator in GP

of S-expressions was chosen to implement GP. The set of functions and terminals that will be used in a specific problem has to be chosen carefully. If the set of functions is not powerful enough, a solution may be very complex or not to be found at all. Like in any evolutionary computation technique, the generation of first population of individuals is important for successful implementation of GP. Some of the other factors that influence the performance of the algorithm are the size of the population, percentage of individuals that participate in the crossover/mutation, maximum depth for the initial individuals and the maximum allowed depth for the generated offspring etc. Some specific advantages of genetic programming are that no analytical knowledge is needed and still could get accurate results. GP approach does scale with the problem size. GP does impose restrictions on how the structure of solutions should be formulated.

1.6 Variants of Genetic Programming

Several variants of GP could be seen in the literature. Some of them are Linear Genetic Programming (LGP), Gene Expression Programming (GEP), Multi Expression Programming (MEP), Cartesian Genetic Programming (CGP), Traceless Genetic Programming (TGP) and Genetic Algorithm for Deriving Software (GADS).

1.6.1 Linear Genetic Programming

Linear genetic programming is a variant of the GP technique that acts on linear genomes [3]. Its main characteristics in comparison to tree-based GP lies in that the evolvable units are not the expressions of a functional programming language (like LISP), but the programs of an imperative language (like c/c ++). This can tremendously hasten the evolution process as, no matter how an individual is initially represented, finally it always has to be represented as a piece of machine code, as fitness evaluation requires physical execution of the individuals. The basic unit of evolution here is a native machine code instruction that runs on the floating-point processor unit (FPU). Since different instructions may have different sizes, here instructions are clubbed up together to form instruction blocks of 32 bits each. The instruction blocks hold one or more native machine code instructions, depending on the sizes of the instructions. A crossover point can occur only between instructions and is prohibited from occurring within an instruction. However the mutation operation does not have any such restriction. LGP uses a specific linear representation of computer programs. A LGP individual is represented by a variable length sequence of simple C language instructions. Instructions operate on one or two indexed variables (registers) r, or on constants c from predefined sets.

An important LGP parameter is the number of registers used by a chromosome. The number of registers is usually equal to the number of attributes of the problem. If the problem has only one attribute, it is impossible to obtain a complex expression such as the quartic polynomial. In that case we have to use several supplementary registers. The number of supplementary registers depends on the complexity of the expression being discovered. An inappropriate choice can have disastrous effects on the program being evolved. LGP uses a modified steady-state algorithm. The initial population is randomly generated. The settings of various linear genetic programming system parameters are of utmost importance for successful performance of the system. The population space has been subdivided into multiple subpopulation or demes. Migration of individuals among the subpopulations causes evolution of the entire population. It helps to maintain diversity in the population, as migration is restricted among the demes. Moreover, the tendency towards a bad local minimum in one deme can be countered by other demes with better search directions. The various LGP search parameters are the mutation frequency, crossover frequency and the reproduction frequency: The crossover operator acts by exchanging sequences of instructions between two tournament winners. Steady state genetic programming approach was used to manage the memory more effectively

1.6.2 Gene Expression Programming (GEP)

The individuals of gene expression programming are encoded in linear chromosomes which are expressed or translated into expression trees (branched entities)[4]. Thus, in GEP, the genotype (the linear chromosomes) and the phenotype (the expression trees) are different entities (both structurally and functionally) that, nevertheless, work together forming an indivisible whole. In contrast to its analogous cellular gene expression, GEP is rather simple. The main players in GEP are only two: the chromosomes and the Expression Trees (ETs), being the latter the expression of the genetic information encoded in the chromosomes. As in nature, the process of information decoding is called translation. And this translation implies obviously a kind of code and a set of rules. The genetic code is very simple: a one-to-one relationship between the symbols of the chromosome and the functions or terminals they represent. The rules are also very simple: they determine the spatial organization of the functions and terminals in the ETs and the type of interaction between sub-ETs. GEP uses linear chromosomes that store expressions in breadth-first form. A GEP gene is a string of terminal and function symbols. GEP genes are composed of a *head* and a *tail*. The head contains both function and terminal symbols. The tail may contain terminal symbols only. For each problem the head length (denoted h) is chosen by the user. The tail length (denoted by t) is evaluated by:

$$t = (n - 1)h + 1, \tag{1.3}$$

where $n$is the number of arguments of the function with more arguments.

GEP genes may be linked by a function symbol in order to obtain a fully functional chromosome. GEP uses mutation, recombination and transposition. GEP uses a generational algorithm. The initial population is randomly generated. The following steps are repeated until a termination criterion is reached: A fixed number of the best individuals enter the next generation (elitism). The mating pool is filled by using binary tournament selection. The individuals from the mating pool are randomly paired and recombined. Two offspring are obtained by recombining two parents. The offspring are mutated and they enter the next generation.

1.6.3 Multi Expression Programming

A GP chromosome generally encodes a single expression (computer program). A Multi Expression Programming (MEP) chromosome encodes several expressions [14]. The best of the encoded solution is chosen to represent the chromosome. The MEP chromosome has some advantages over the single-expression chromosome especially when the complexity of the target expression is not known. This feature also acts as a provider of variable-length expressions. MEP genes are represented by substrings of a variable length. The number of genes per chromosome is constant. This number defines the length of the chromosome. Each gene encodes a terminal or a function symbol. A gene that encodes a function includes pointers towards the function arguments. Function arguments always have indices of lower values than the position of the function itself in the chromosome.

The proposed representation ensures that no cycle arises while the chromosome is decoded (phenotypically transcripted). According to the proposed representation scheme, the first symbol of the chromosome must be a terminal symbol. In this way, only syntactically correct programs (MEP individuals) are obtained. The maximum number of symbols in MEP chromosome is given by the formula:

$$Number\_of\_Symbols = (n + 1) \times (Number\_of\_Genes - 1) + 1, \qquad (1.4)$$

where n is the number of arguments of the function with the greatest number of arguments. The translation of a MEP chromosome into a computer program represents the phenotypic transcription of the MEP chromosomes. Phenotypic translation is obtained by parsing the chromosome top-down. A terminal symbol specifies a simple expression. A function symbol specifies a complex expression obtained by connecting the operands specified by the argument positions with the current function symbol.

Due to its multi expression representation, each MEP chromosome may be viewed as a forest of trees rather than as a single tree, which is the case of Genetic Programming.

1.6.4 Cartesian Genetic Programming

Cartesian Genetic Programming (CGP) uses a network of nodes (indexed graph) to achieve an input to output mapping [13]. Each node consists of a number of inputs, these being used as parameters in a determined mathematical or logical function to create the node output. The functionality and connectivity of the nodes are stored as a string of numbers (the genotype) and evolved to achieve the optimum mapping. The genotype is then mapped to an indexed graph that can be executed as a program.

In CGP there are very large number of genotypes that map to identical genotypes due to the presence of a large amount of redundancy. Firstly there is node redundancy that is caused by genes associated with nodes that are not part of the connected graph representing the program. Another form of redundancy in CGP, also present in all other forms of GP is, functional redundancy.

1.6.5 Traceless Genetic Programming (TGP)

The main difference between Traceless Genetic Programming and GP is that TGP does not explicitly store the evolved computer programs [15]. TGP is useful when the trace (the way in which the results are obtained) between the input and output is not important. TGP uses two genetic operators: crossover and insertion. The insertion operator is useful when the population contains individuals representing very complex expressions that cannot improve the search.

1.6.6 Grammatical Evolution

Grammatical evolution [18] is a grammar-based, linear genome system. In grammatical evolution, the Backus Naur Form (BNF) specification of a language is used to describe the output produced by the system (a compilable code fragment). Different BNF grammars can be used to produce code automatically in any language. The genotype is a string of eight-bit binary numbers generated at random and treated as integer values from 0 to 255. The phenotype is a running computer program generated by a genotype-phenotype mapping process. The genotype-phenotype mapping in grammatical evolution is deterministic because each individual is always mapped to the same phenotype. In grammatical evolution, standard genetic algorithms are applied to the different genotypes in a population using the typical crossover and mutation operators.

1.6.7 Genetic Algorithm for Deriving Software (GADS)

Genetic algorithm for deriving software is a GP technique where the genotype is distinct from the phenotype [16]. The GADS genotype is a list of integers representing productions in a syntax. This is used to generate the phenotype, which is a program in the language defined by the syntax. Syntactically invalid phenotypes cannot be generated, though there may be phenotypes with residual nonterminals.

1.7 Summary

This chapter presented the biological motivation and fundamental aspects of evolutionary algorithms and its constituents, namely genetic algorithm, evolution strategies, evolutionary programming and genetic programming. Most popular variants of genetic programming are introduced. Important advantages of evolutionary computation while compared to classical optimization techniques are also discussed.

References

1. Abraham, A., Evolutionary Computation, In: Handbook for Measurement, Systems Design, Peter Sydenham and Richard Thorn (Eds.), John Wiley and Sons Ltd., London, ISBN 0-470-02143-8, pp. 920–931, 2005.
2. Bäck, T., Evolutionary algorithms in theory and practice: Evolution Strategies, Evolutionary Programming, Genetic Algorithms, Oxford University Press, New York, 1996.
3. Banzhaf, W., Nordin, P., Keller, E. R., Francone, F. D., Genetic Programming : An Introduction on The Automatic Evolution of Computer Programs and its Applications, Morgan Kaufmann Publishers, Inc., 1998.

4. Ferreira, C., Gene Expression Programming: A new adaptive algorithm for solving problems - Complex Systems, Vol. 13, No. 2, pp. 87–129, 2001.
5. Fogel, L.J., Owens, A.J. and Walsh, M.J., Artificial Intelligence Through Simulated Evolution, John Wiley & Sons Inc. USA, 1966.
6. Fogel, D. B. (1999) Evolutionary Computation: Toward a New Philosophy of Machine Intelligence. IEEE Press, Piscataway, NJ, Second edition, 1999.
7. Goldberg, D. E., Genetic Algorithms in search, optimization, and machine learning, Reading: Addison-Wesley Publishing Corporation Inc., 1989.
8. History of Lisp, `http://www-formal.stanford.edu/jmc/history/lisp.html`, 2004.
9. Holland, J. Adaptation in Natural and Artificial Systems, Ann Harbor: University of Michican Press, 1975.
10. Jang, J.S.R., Sun, C.T. and Mizutani, E., Neuro-Fuzzy and Soft Computing: A Computational Approach to Learning and Machine Intelligence, Prentice Hall Inc, USA, 1997.
11. Koza. J. R., Genetic Programming. The MIT Press, Cambridge, Massachusetts, 1992.
12. Michalewicz, Z., Genetic Algorithms + Data Structures = Evolution Programs, Berlin: Springer-Verlag, 1992.
13. Miller, J. F. Thomson, P., Cartesian Genetic Programming, Proceedings of the European Conference on Genetic Programming, Lecture Notes In Computer Science, Vol. 1802 pp. 121–132, 2000.
14. Oltean M. and Grosan C., Evolving Evolutionary Algorithms using Multi Expression Programming. Proceedings of The 7th. European Conference on Artificial Life, Dortmund, Germany, pp. 651–658, 2003.
15. Oltean, M., Solving Even-Parity Problems using Traceless Genetic Programming, IEEE Congress on Evolutionary Computation, Portland, G. Greenwood, et. al (Eds.), IEEE Press, pp. 1813–1819, 2004.
16. Paterson, N. R. and Livesey, M., Distinguishing Genotype and Phenotype in Genetic Programming, Late Breaking Papers at the Genetic Programming 1996, J. R. Koza (Ed.), pp. 141–150,1996.
17. Rechenberg, I., Evolutionsstrategie: Optimierung technischer Systeme nach Prinzipien der biologischen Evolution, Stuttgart: Fromman-Holzboog, 1973.
18. Ryan, C., Collins, J. J. and O'Neill, M., Grammatical Evolution: Evolving Programs for an Arbitrary Language, Proceedings of the First European Workshop on Genetic Programming (EuroGP'98), Lecture Notes in Computer Science 1391, pp. 83-95, 1998.
19. Schwefel, H.P., Numerische Optimierung von Computermodellen mittels der Evolutionsstrategie, Basel: Birkhaeuser, 1977.
20. Törn A. and Zilinskas A., Global Optimization, Lecture Notes in Computer Science, Vol. 350, Springer-Verlag, Berlin, 1989.

2

Automatically Defined Functions in Gene Expression Programming

Cândida Ferreira

Gepsoft, 73 Elmtree Drive, Bristol BS13 8NA,
United Kingdom
candidaf@gepsoft.com,
http://www.gene-expression-programming.com/gep/author.asp

In this chapter it is shown how Automatically Defined Functions are encoded in the genotype/phenotype system of Gene Expression Programming. As an introduction, the fundamental differences between Gene Expression Programming and its predecessors, Genetic Algorithms and Genetic Programming, are briefly summarized so that the evolutionary advantages of Gene Expression Programming are better understood. The introduction proceeds with a detailed description of the architecture of the main players of Gene Expression Programming (chromosomes and expression trees), focusing mainly on the interactions between them and how the simple, yet revolutionary, structure of the chromosomes allows the efficient, unconstrained exploration of the search space. The work proceeds with an introduction to Automatically Defined Functions and how they are implemented in Gene Expression Programming. Furthermore, the importance of Automatically Defined Functions in Evolutionary Computation is thoroughly analyzed by comparing the performance of sophisticated learning systems with Automatically Defined Functions with much simpler ones on the sextic polynomial problem.

2.1 Genetic Algorithms: Historical Background

The way nature solves problems and creates complexity has inspired scientists to create artificial systems that learn how to solve a particular problem without human intervention. The first attempts were done in the 1950s by Friedberg [9, 10], but ever since highly sophisticated systems have been developed that apply Darwin's ideas of natural evolution to the artificial world of computers and modeling. Of particular interest to this work are the Genetic Algorithms (GAs) and the Genetic Programming (GP) technique as they are

C. Ferreira: *Automatically Defined Functions in Gene Expression Programming*, Studies in Computational Intelligence (SCI) **13**, 21–56 (2006)
www.springerlink.com

the predecessors of Gene Expression Programming (GEP), an extremely versatile genotype/phenotype system. The way Automatically Defined Functions (ADFs) are implemented in GEP is another example of the great versatility of this algorithm and the versatility of GEP ADFs opens up new grounds for the creation of even more sophisticated artificial learning systems. So let's start by introducing briefly these three techniques in order to appreciate the versatility of the genotype/phenotype system of Gene Expression Programming with and without ADFs.

2.1.1 Genetic Algorithms

Genetic Algorithms were invented by John Holland in the 1960s and they also apply biological evolution theory to computer systems [11]. And like all evolutionary computer systems, GAs are an oversimplification of biological evolution. In this case, solutions to a problem are usually encoded in fixed length strings of 0's and 1's (chromosomes), and populations of such strings (individuals or candidate solutions) are manipulated in order to evolve a good solution to a particular problem. From generation to generation individuals are reproduced with modification and selected according to fitness. Modification in the original genetic algorithm was introduced by the search operators of mutation, crossover, and inversion, but more recent applications started favoring mutation and crossover, dropping inversion in the process.

It is worth pointing out that GAs' individuals consist of naked chromosomes or, in other words, GAs' individuals are simple replicators. And like all simple replicators, the chromosomes of GAs work both as genotype and phenotype. This means that they are simultaneously the objects of selection and the guardians of the genetic information that must be replicated and passed on with modification to the next generation. Consequently, the whole structure of the replicator determines the functionality and, consequently, the fitness of the individual. For instance, in such systems it is not possible to use only a particular region of the replicator as a solution to a problem; the whole replicator is always the solution: nothing more, nothing less.

2.1.2 Genetic Programming

Genetic Programming, invented by Cramer in 1985 [1] and further developed by Koza [14], finds an alternative to fixed length solutions through the introduction of nonlinear structures (parse trees) with different sizes and shapes. The alphabet used to create these structures is also more varied than the 0's and 1's of GAs' individuals, creating a richer, more versatile system of representation. Notwithstanding, GP individuals also lack a simple, autonomous genome: like the linear chromosomes of GAs, the nonlinear structures of GP are also naked replicators cursed with the dual role of genotype/phenotype.

It is worth noticing that the parse trees of GP resemble protein molecules in their use of a richer alphabet and in their complex and unique hierarchical representation. Indeed, parse trees are capable of exhibiting a great variety of functionalities. The problem with these complex replicators is that their reproduction with modification is highly constrained in evolutionary terms, simply because the modifications must take place on the parse tree itself and, consequently, only a limited range of modification is possible. Indeed, the genetic operators of GP operate at the tree level, modifying or exchanging particular branches between trees.

Although at first sight this might appear advantageous, it greatly limits the GP technique (we all know the limits of grafting and pruning in nature). Consider, for instance, crossover, the most used and often the only search operator used in GP (Figure 2.1). In this case, selected branches are exchanged between two parent trees to create offspring. The idea behind its implementation was to exchange smaller, mathematically concise blocks in order to evolve more complex, hierarchical solutions composed of simpler building blocks, guaranteeing, at the same time, the creation of syntactically correct structures.

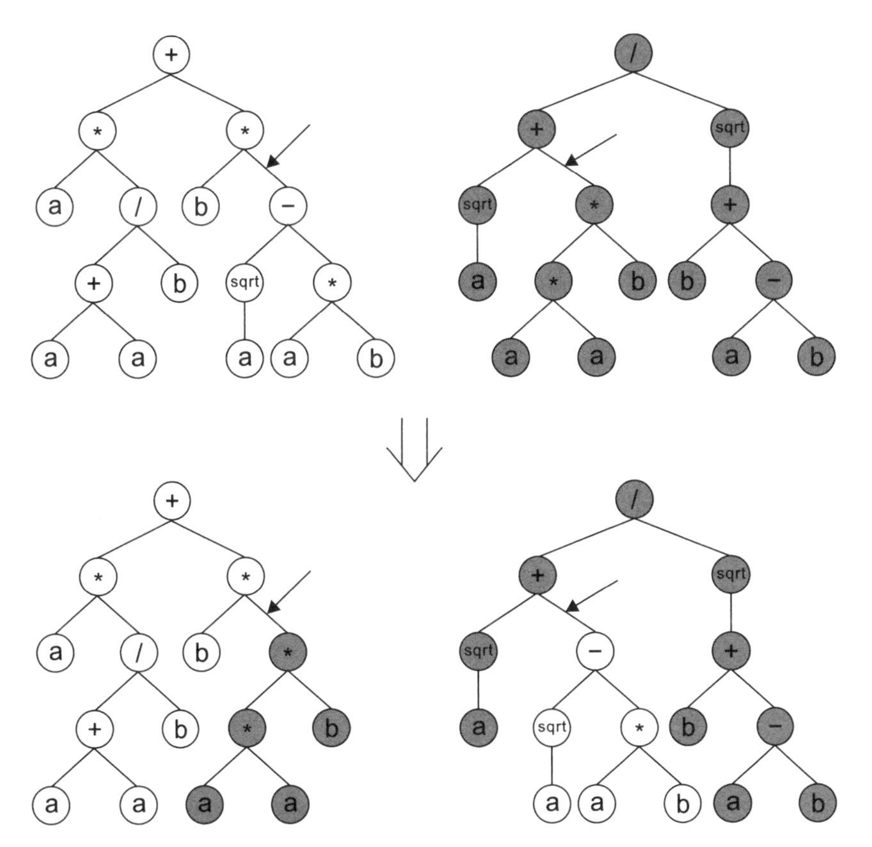

Fig. 2.1. Tree crossover in Genetic Programming. The arrows indicate the crossover points.

The mutation operator in GP is also very different from natural point mutation. This operator selects a node in the parse tree and replaces the branch underneath by a new randomly generated branch (Figure 2.2). Notice that the overall shape of the tree is not greatly changed by this kind of mutation, especially if lower nodes are preferentially chosen as mutation targets.

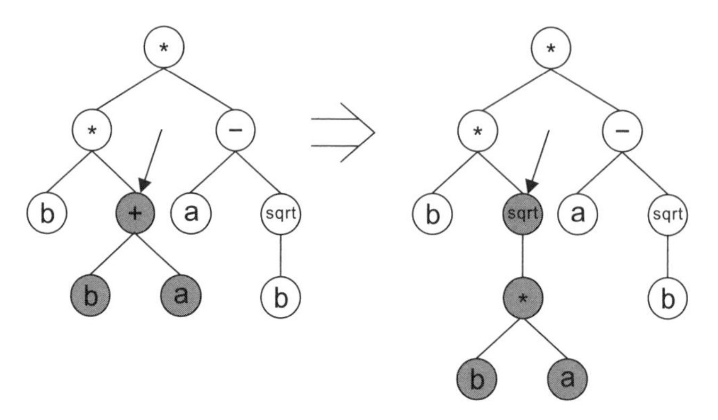

Fig. 2.2. Tree mutation in Genetic Programming. The arrow indicates the mutation point. The new branch randomly generated by the mutation operator in the daughter tree is shown in gray.

Permutation is the third operator used in Genetic Programming and the most conservative of the three. During permutation, the arguments of a randomly chosen function are randomly permuted (Figure 2.3). In this case the overall shape of the tree remains unchanged.

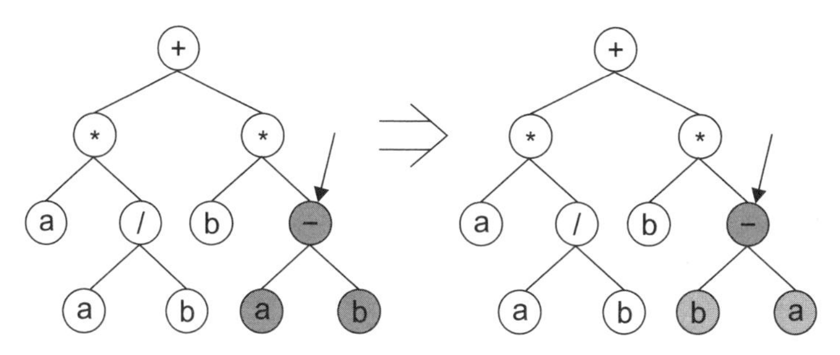

Fig. 2.3. Permutation in Genetic Programming. The arrow indicates the permutation point. Note that the arguments of the permuted function traded places in the daughter tree.

In summary, in Genetic Programming the operators resemble more of a conscious mathematician than the blind way of nature. But in adaptive systems the blind way of nature is much more efficient and systems such as GP are highly limited in evolutionary terms. For instance, the implementation of other operators in GP, such as the simple yet high-performing point mutation [6], is unproductive as most mutations would have resulted in syntactically incorrect structures (Figure 2.4). Obviously, the implementation of other operators such as transposition or inversion raises similar difficulties and the search space in GP remains vastly unexplored.

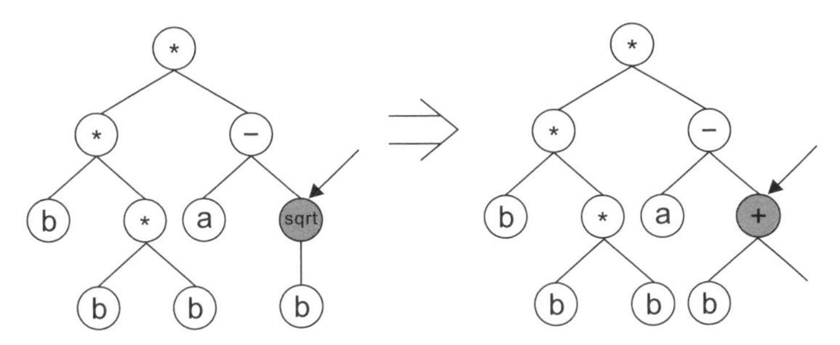

Fig. 2.4. Illustration of a hypothetical event of point mutation in Genetic Programming. The arrow indicates the mutation point. Note that the daughter tree is an invalid structure.

Although Koza described these three genetic operators as the basic GP operators, crossover is practically the only search operator used in most GP applications [13, 14, 15]. Consequently, no new genetic material is introduced in the genetic pool of GP populations. Not surprisingly, huge populations of parse trees must be used in order to prime the initial population with all the necessary building blocks so that good solutions could be discovered by just moving these initial building blocks around.

Finally, due to the dual role of the parse trees (genotype and phenotype), Genetic Programming is also incapable of a simple, rudimentary expression; in all cases, the entire parse tree is the solution: nothing more, nothing less.

2.1.3 Gene Expression Programming

Gene Expression Programming was invented by myself in 1999 [3], and incorporates both the simple, linear chromosomes of fixed length similar to the ones used in Genetic Algorithms and the ramified structures of different sizes and shapes similar to the parse trees of Genetic Programming. And since the ramified structures of different sizes and shapes are totally encoded in the linear chromosomes of fixed length, this is equivalent to say that, in GEP, the genotype and phenotype are finally separated from one another and the system can now benefit from all the evolutionary advantages this brings about.

Thus, the phenotype of GEP consists of the same kind of ramified structure used in GP. But the ramified structures evolved by GEP (called expression trees) are the expression of a totally autonomous genome. Therefore, with

GEP, a remarkable thing happened: the second evolutionary threshold – the phenotype threshold – was crossed [2]. And this means that only the genome (slightly modified) is passed on to the next generation. Consequently, one no longer needs to replicate and mutate rather cumbersome structures as all the modifications take place in a simple linear structure which only later will grow into an expression tree.

The fundamental steps of Gene Expression Programming are schematically represented in Figure 2.5. The process begins with the random generation of the chromosomes of a certain number of individuals (the initial population). Then these chromosomes are expressed and the fitness of each individual is evaluated against a set of fitness cases (also called selection environment). The individuals are then selected according to their fitness (their performance in that particular environment) to reproduce with modification, leaving progeny with new traits. These new individuals are, in their turn, subjected to the same developmental process: expression of the genomes, confrontation of the selection environment, selection according to fitness, and reproduction with modification. The process is repeated for a certain number of generations or until a good solution has been found.

So, the pivotal insight of Gene Expression Programming consisted in the invention of chromosomes capable of representing any parse tree. For that purpose a new language – Karva language – was created in order to read and express the information encoded in the chromosomes. The details of this new language are given in the next section.

Furthermore, the structure of the chromosomes was designed in order to allow the creation of multiple genes, each coding for a smaller program or sub-expression tree. It is worth emphasizing that Gene Expression Programming is the only genetic algorithm with multiple genes. Indeed, in truly functional genotype/phenotype systems, the creation of more complex individuals composed of multiple genes is a child's play, and illustrates quite well the great versatility of the GEP system. In fact, after their inception, these systems seem to catapult themselves into higher levels of complexity such as the uni- and multicellular systems, where different cells put together different combinations of genes [4]. We will see later in this chapter how the cellular system of GEP is an extremely elegant way of implementing Automatically Defined Functions that may be reused by the created programs.

The basis for all this novelty resides on the simple, yet revolutionary structure of GEP genes. This structure not only allows the encoding of any conceivable program but also allows an efficient evolution. This versatile structural organization also allows the implementation of a very powerful set of genetic operators which can then very efficiently search the solution space. As in nature, the search operators of GEP always generate valid structures and therefore are remarkably suited to creating genetic diversity.

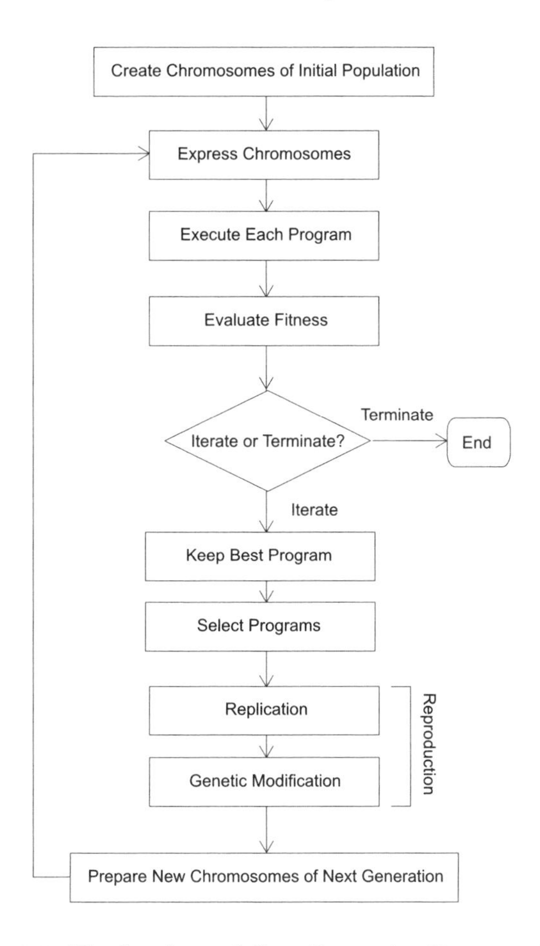

Fig. 2.5. The flowchart of Gene Expression Programming.

2.2 The Architecture of GEP Individuals

We know already that the main players in Gene Expression Programming are the chromosomes and the expression trees (ETs), and that the latter are the expression of the genetic information encoded in the former. As in nature, the process of information decoding is called translation. And this translation implies obviously a kind of code and a set of rules. The genetic code is very simple: a one-to-one relationship between the symbols of the chromosome and the nodes they represent in the trees. The rules are also very simple: they determine the spatial organization of nodes in the expression trees and the type of interaction between sub-ETs. Therefore, there are two languages in GEP: the language of the genes and the language of expression trees and, thanks to the simple rules that determine the structure of ETs and their interactions, we will see that it is possible to infer immediately the phenotype given the sequence of a gene, and vice versa. This means that we can choose to have a very complex program represented by its compact genome without losing any information. This unequivocal bilingual notation is called Karva language. Its details are explained in the remainder of this section.

2.2.1 Open Reading Frames and Genes

The structural organization of GEP genes is better understood in terms of open reading frames (ORFs). In biology, an ORF or coding sequence of a gene begins with the start codon, continues with the amino acid codons, and ends at a termination codon. However, a gene is more than the respective ORF, with sequences upstream of the start codon and sequences downstream of the stop codon. Although in GEP the start site is always the first position of a gene, the termination point does not always coincide with the last position of a gene. Consequently, it is common for GEP genes to have noncoding regions downstream of the termination point. (For now we will not consider these noncoding regions, as they do not interfere with expression.)

Consider, for example, the algebraic expression:

$$\frac{\sqrt{a+b}}{c-d} \tag{2.1}$$

It can also be represented as a diagram or ET:

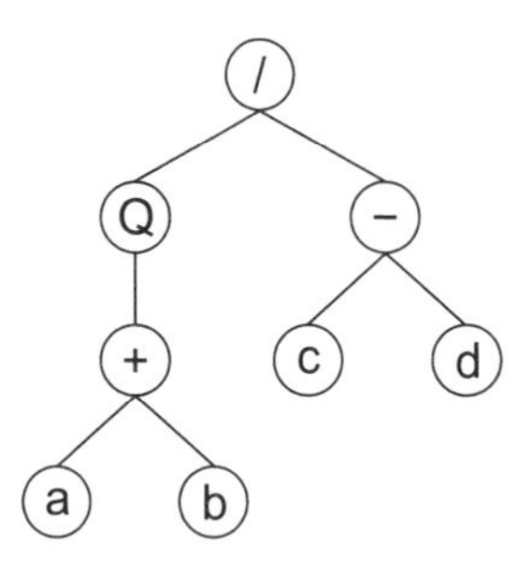

where "Q" represents the square root function.

This kind of diagram representation is in fact the phenotype of GEP chromosomes. And the genotype can be easily inferred from the phenotype as follows:

```
01234567
/Q-+cdab
```
(2.2)

which is the straightforward reading of the ET from left to right and from top to bottom (exactly as one reads a page of text). The expression (2.2) is an open reading frame, starting at "/" (position 0) and terminating at "b" (position 7). These ORFs were named *K-expressions* from Karva language.

Consider another ORF, the following K-expression:

```
0123456789
*//aQ*bddc
```
(2.3)

Its expression as an ET is also very simple and straightforward. In order to express the ORF correctly, we must follow the rules governing the spatial distribution of functions and terminals. First, the start of a gene corresponds to the root of the expression tree, and it occupies the topmost position (or first line) on the tree. Second, in the next line, below each function, are placed as many branch nodes as there are arguments to that function. Third, from left to right, the nodes are filled consecutively with the next elements of the K-expression. Fourth, the process is repeated until a line containing only terminals is formed. In this case, the expression tree is formed:

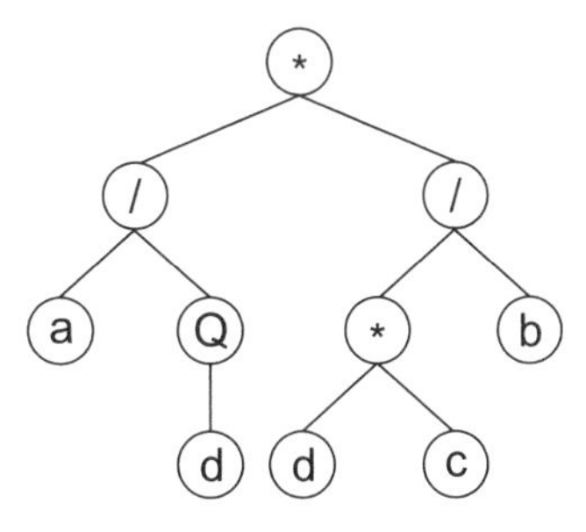

Looking at the structure of ORFs only, it is difficult or even impossible to see the advantages of such a representation, except perhaps for its simplicity and elegance. However, when open reading frames are analyzed in the context of a gene, the advantages of this representation become obvious. As I said before, GEP chromosomes have fixed length, and they are composed of one or more genes of equal length. Consequently, the length of a gene is also fixed. Thus, in GEP, what changes is not the length of genes, but rather the length of the ORF. Indeed, the length of an ORF may be equal to or less than the length of the gene. In the first case, the termination point coincides with the end of the gene, and in the latter, the termination point is somewhere upstream of the end of the gene. And this obviously means that GEP genes have, most of the times, noncoding regions at their ends.

And what is the function of these noncoding regions at the end of GEP genes? We will see that they are the essence of GEP and evolvability, because they allow the modification of the genome using all kinds of genetic operators without any kind of restriction, always producing syntactically correct programs. Thus, in GEP, the fundamental property of genotype/phenotype systems – syntactic closure – is intrinsic, allowing the totally unconstrained restructuring of the genotype and, consequently, an efficient evolution.

In the next section we are going to analyze the structural organization of GEP genes in order to understand how they invariably code for syntactically correct programs and why they allow the unconstrained application of virtually any genetic operator.

2.2.2 Structural Organization of Genes

The novelty of GEP genes resides in the fact that they are composed of a head and a tail. The head contains symbols that represent both functions and terminals, whereas the tail contains only terminals. For each problem, the length of the head h is chosen, whereas the length of the tail t is a function of h and the number of arguments n of the function with more arguments (also called maximum arity) and is evaluated by the equation:

$$t = h(n-1) + 1 \tag{2.4}$$

Consider a gene for which the set of terminals $T = \{\text{a, b}\}$ and the set of functions $F = \{\text{Q}, *, /, \text{-}, +\}$, thus giving $n = 2$. And if we chose a head lenght $h = 10$, then $t = 10\ (2 - 1) + 1 = 11$ and the length of the gene g is $10 + 11 = 21$. One such gene is shown below (the tail is underlined):

$$\begin{array}{l}\texttt{012345678901234567890}\\ \texttt{+*/a-Qbb/*}\underline{\texttt{aabaabbabaa}}\end{array} \tag{2.5}$$

It codes for the following expression tree:

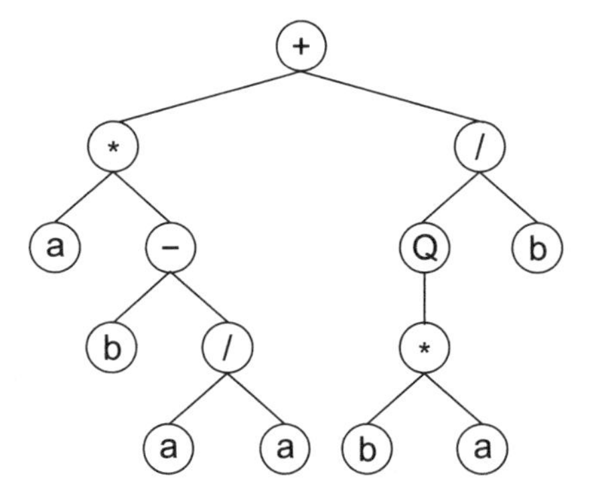

Note that, in this case, the open reading frame ends at position 13, whereas the gene ends at position 20.

Suppose now a mutation occurred at position 3, changing the "a" into "+". Then the following gene is obtained:

$$\begin{array}{l}\texttt{012345678901234567890}\\ \texttt{+*/}\underline{\texttt{+}}\texttt{-Qbb/*aabaabbabaa}\end{array} \tag{2.6}$$

And its expression gives:

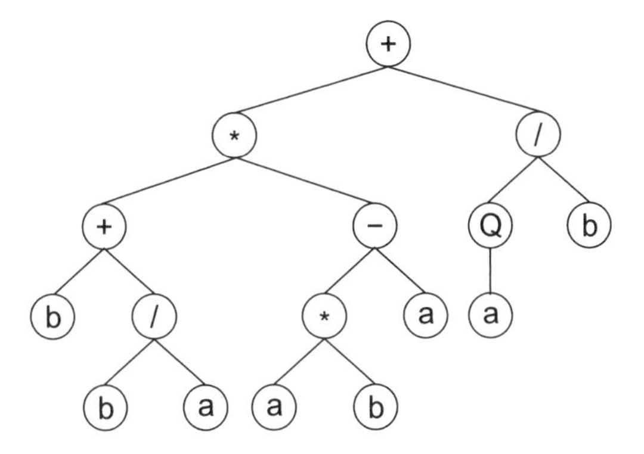

In this case, the termination point shifts two positions to the right (position 15), enlarging and changing significantly the daughter tree.

Obviously the opposite might also happen, and the daughter tree might shrink. For example, consider again gene (2.5) above, and suppose a mutation occurred at position 2, changing the "/" into "b":

```
012345678901234567890
++ba-Qbb/*aabaabbabaa
```
(2.7)

And now its expression results in the following ET:

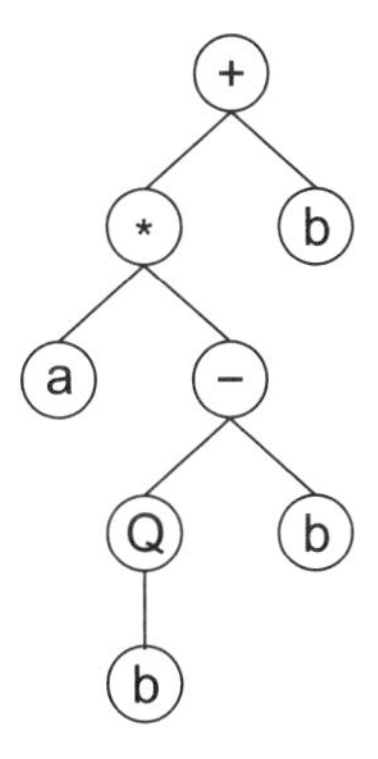

In this case, the ORF ends at position 7, shortening the original ET by six nodes.

So, despite their fixed length, the genes of Gene Expression Programming have the potential to code for expression trees of different sizes and shapes, where the simplest is composed of only one node (when the first element of a gene is a terminal) and the largest is composed of as many nodes as the length of the gene (when all the elements of the head are functions with maximum arity).

It is evident from the examples above, that any modification made in the genome, no matter how profound, always results in a structurally correct program. Consequently, the implementation of a powerful set of search operators, such as point mutation, inversion, transposition, and recombination, is a child's play, making Gene Expression Programming the ideal playground for the discovery of the perfect solution using an economy of resources (see [3] and [7] for a detailed description of the mechanisms and effects of the different genetic operators commonly used in Gene Expression Programming).

2.2.3 Multigenic Chromosomes and Linking Functions

The chromosomes of Gene Expression Programming are usually composed of more than one gene of equal length. For each problem or run, the number of genes, as well as the length of the head, are a priori chosen. Each gene codes for a sub-ET and, in problems with just one output, the sub-ETs interact with one another forming a more complex multi-subunit ET; in problems with multiple outputs, though, each sub-ET evolves its respective output.

Consider, for example, the following chromosome with length 39, composed of three genes, each with length 13 (the tails are underlined):

```
012345678901201234567890120123456789012
*Q-b/abbbaaba/aQb-bbbaabaa*Q-/b*abbbbaa        (2.8)
```

It has three open reading frames, and each ORF codes for a sub-ET (Figure 2.6). We know already that the start of each ORF coincides with the first element of the gene and, for the sake of clarity, for each gene it is always indicated by position zero; the end of each ORF, though, is only evident upon construction of the corresponding sub-ET. As you can see in Figure 2.6, the first open reading frame ends at position 7; the second ORF ends at position 3; and the last ORF ends at position 9. Thus, GEP chromosomes contain several ORFs of different sizes, each ORF coding for a structurally and functionally unique sub-ET. Depending on the problem at hand, these sub-ETs may be selected individually according to their respective outputs, or they may form a more complex, multi-subunit expression tree and be selected as a whole. In these multi-subunit structures, individual sub-ETs interact with one another by a particular kind of posttranslational interaction or linking. For instance, algebraic sub-ETs can be linked by addition or subtraction whereas Boolean sub-ETs can be linked by OR or AND.

The linking of three sub-ETs by addition is illustrated in Figure 2.6, c. Note that the final ET could be linearly encoded as the following K-expression:

```
012345678901234567890123
++**/Q-Q-aQ/b*b/ababbbbb        (2.9)
```

However, the use of multigenic chromosomes is more appropriate to evolve good solutions to complex problems, for they permit the modular construction of complex, hierarchical structures, where each gene codes for a smaller and simpler building block. These building blocks are physically separated from one another, and thus can evolve independently. Not surprisingly, these multigenic systems are much more efficient than unigenic ones [3, 4]. And, of

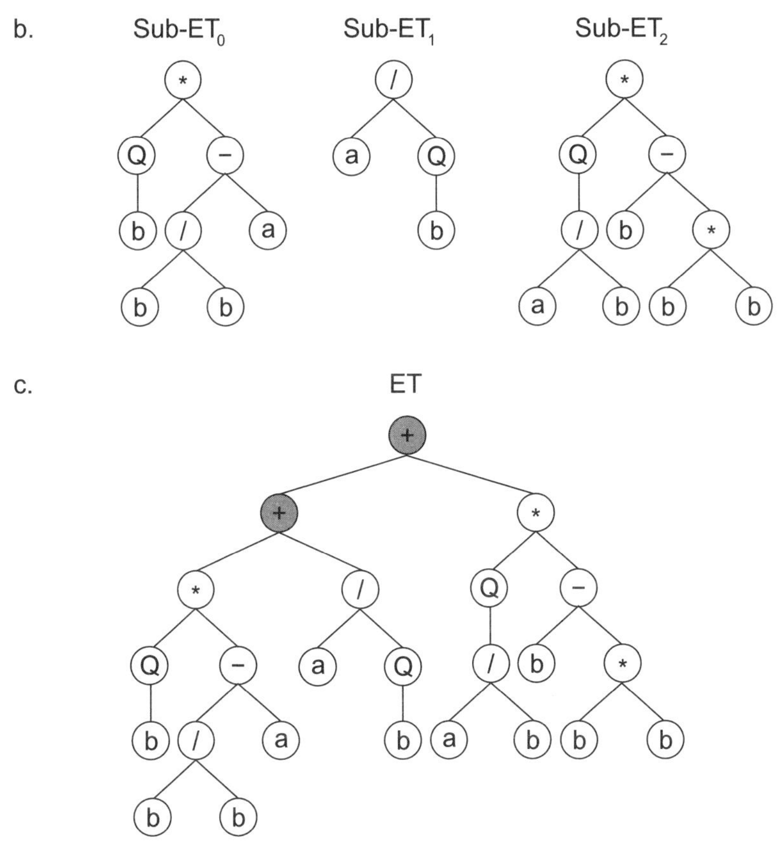

Fig. 2.6. Expression of GEP genes as sub-ETs. **a)** A three-genic chromosome with the tails shown in bold. Position zero marks the start of each gene. **b)** The sub-ETs codified by each gene. **c)** The result of posttranlational linking with addition. The linking functions are shown in gray.

course, they also open up new grounds to solve problems of multiple outputs, such as parameter optimization or classification problems [4].

2.3 Chromosome Domains and Random Numerical Constants

We have already met two different domains in GEP genes: the head and the tail. And now another one – the Dc domain – will be introduced. This domain was especially designed to handle random numerical constants and consists of an extremely elegant, efficient, and original way of dealing with them.

As an example, Genetic Programming handles random constants by using a special terminal named "ephemeral random constant" [14]. For each ephemeral random constant used in the trees of the initial population, a random number of a special data type in a specified range is generated. Then these random constants are moved around from tree to tree by the crossover operator. Note, however, that with this method no new constants are created during evolution.

Gene Expression Programming handles random numerical constants differently [3]. GEP uses an extra terminal "?" and an extra domain Dc composed of the symbols chosen to represent the random constants. The values of each random constant, though, are only assigned during gene expression. Thus, for each gene, the initial random constants are generated during the creation of the initial population and kept in an array. However, a special operator is used to introduce genetic variation in the available pool of random constants by mutating the random constants directly. In addition, the usual operators of GEP (mutation, inversion, transposition, and recombination), plus a Dc-specific transposition and a Dc-specific inversion, guarantee the effective circulation of the random constants in the population. Indeed, with this scheme of constants' manipulation, the appropriate diversity of numerical constants is generated at the beginning of a run and maintained easily afterwards by the genetic operators. Let's take then a closer look at the structural organization of this extra domain and how it interacts with the sub-ET encoded in the head/tail structure.

Structurally, the Dc comes after the tail, has a length equal to t, and is composed of the symbols used to represent the random constants. Therefore, another region with defined boundaries and its own alphabet is created in the gene.

Consider the single-gene chromosome with an $h = 8$ (the Dc is underlined):

```
01234567890123456789012345
*-Q*Q*?+?ba?babba238024198
```
(2.10)

where the terminal "?" represents the random constants. The expression of this kind of chromosome is done exactly as before, obtaining:

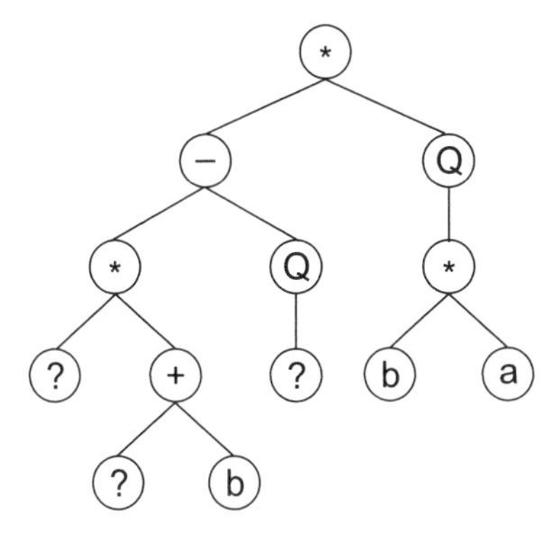

Then the ?'s in the expression tree are replaced from left to right and from top to bottom by the symbols (numerals, for simplicity) in Dc, obtaining:

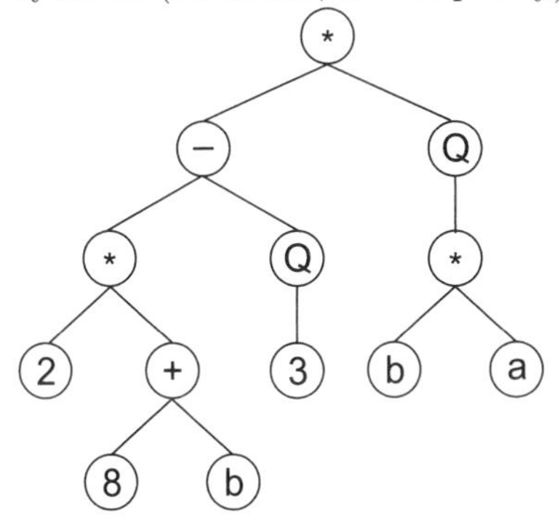

The random constants corresponding to these symbols are kept in an array and, for simplicity, the number represented by the numeral indicates the order in the array. For instance, for the following array of 10 elements:

$$A = \{1.095, 1.816, 2.399, 1.618, 0.725, 1.997, 0.094, 2.998, 2.826, 2.057\}$$

the chromosome (2.10) above gives:

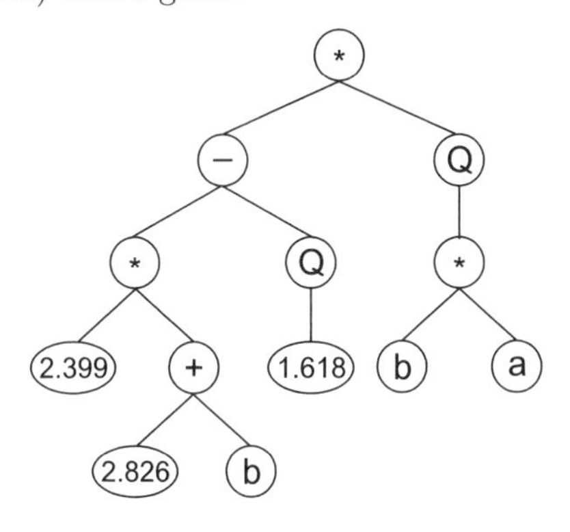

This type of domain was used to great advantage not only in symbolic regression but also in parameter optimization and polynomial induction [4]. But this elegant structure can also be used to evolve the weights and thresholds of neural networks [8] and to encode decision trees with numerical attributes (unpublished material). And we will see here for the first time that this kind of domain can also be used to create Automatically Defined Functions with random numerical constants.

2.4 Cells and the Creation of Automatically Defined Functions

Automatically Defined Functions were for the first time introduced by Koza as a way of reusing code in Genetic Programming [14].

The ADFs of GP obey a rigid syntax in which an S-expression, with a LIST-n function on the root, lists n-1 function-defining branches and one value-returning branch (Figure 2.7). The function-defining branches are used to create ADFs that may or may not be called upon by the value-returning branch. Such rigid structure imposes great constraints on the genetic operators as the different branches of the LIST function are not allowed to exchange genetic material amongst themselves. Furthermore, the ADFs of Genetic Programming are further constrained by the number of arguments each takes, as the number of arguments must be a priori defined and cannot be changed during evolution.

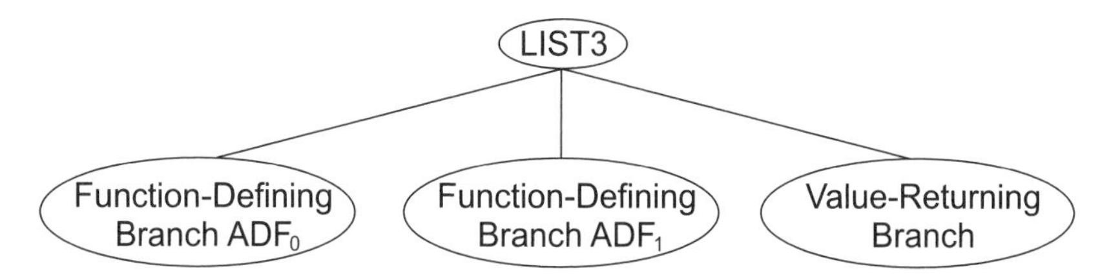

Fig. 2.7. The overall structure of an S-expression with two function-defining branches and the value-returning branch used in Genetic Programming to create Automatically Defined Functions.

In the multigenic system of Gene Expression Programming, the implementation of Automatically Defined Functions can be done elegantly and without any kind of constraints as each gene is used to encode a different ADF [4]. The way these ADFs interact with one another and how often they are called upon is encoded in special genes – homeotic genes – thus called because they are the ones controlling the overall development of the individual. And continuing with the biological analogy, the product of expression of such genes is also called a cell. Thus, homeotic genes determine which genes are expressed in which cell and how they interact with one another. Or stated differently, homeotic genes determine which ADFs are called upon in which main program and how they interact with one another. How this is done is explained in the remainder of this section.

2.4.1 Homeotic Genes and the Cellular System of GEP

Homeotic genes have exactly the same kind of structure as conventional genes and are built using an identical process. They also contain a head and a tail domain, with the heads containing, in this case, linking functions (so called because they are actually used to link different ADFs) and a special class of terminals – genic terminals – representing conventional genes, which, in the cellular system, encode different ADFs; the tails contain obviously only genic terminals.

Consider, for instance, the following chromosome:

```
0123456780123456780123456780123456789O
/-b/abbaa*a-/abbab-*+abbbaa**Q2+010102
```
(2.11)

It codes for three conventional genes and one homeotic gene (underlined). The conventional genes encode, as usual, three different sub-ETs, with the difference that now these sub-ETs will act as ADFs and, therefore, may be invoked multiple times from different places. And the homeotic gene controls the interactions between the different ADFs (Figure 2.8). As you can see in Figure 2.8, in this particular case, ADF_0 is used twice in the main program, whereas ADF_1 and ADF_2 are both used just once.

It is worth pointing out that homeotic genes have their specific length and their specific set of functions. And these functions can take any number of arguments (functions with one, two, three, ..., n, arguments). For instance, in the particular case of chromosome (2.11), the head length of the homeotic gene h_H is equal to five, whereas for the conventional genes $h = 4$; the function set used in the homeotic gene F_H consists of $F_H = \{+, -, *, /, Q\}$, whereas for the conventional genes the function set consists of $F = \{+, -, *, /\}$. As shown in Figure 2.8, this cellular system is not only a form of elegantly allowing the evolution of linking functions in multigenic systems but also an extremely elegant way of encoding ADFs that can be called an arbitrary number of times from an arbitrary number of different places.

2.4.2 Multicellular Systems

The use of more than one homeotic gene results obviously in a multicellular system, in which each homeotic gene puts together a different consortium of genes.

Consider, for instance, the following chromosome:

```
0123456012345601234560123456780123456789
*Q-bbabQ*baabb-/abbab*+21Q1102/*21+1011
```
(2.12)

It codes for three conventional genes and two homeotic genes (underlined). And its expression results in two different cells or programs, each expressing different genes in different ways (Figure 2.9). As you can see in Figure 2.9, ADF_1 is used twice in both cells; ADF_2 is used just once in both cells; and ADF_0 is only used in $Cell_1$.

The applications of these multicellular systems are multiple and varied and, like the multigenic systems, they can be used both in problems with just one output and in problems with multiple outputs. In the former case, the best program or cell accounts for the fitness of the individual; in the latter, each cell is responsible for a particular facet in a multiple output task such as a classification task with multiple classes.

a.
```
012345678012345678012345678012345678 90
/-b/abbaa*a-/abbab-*+abbbaa**Q2+010102
```

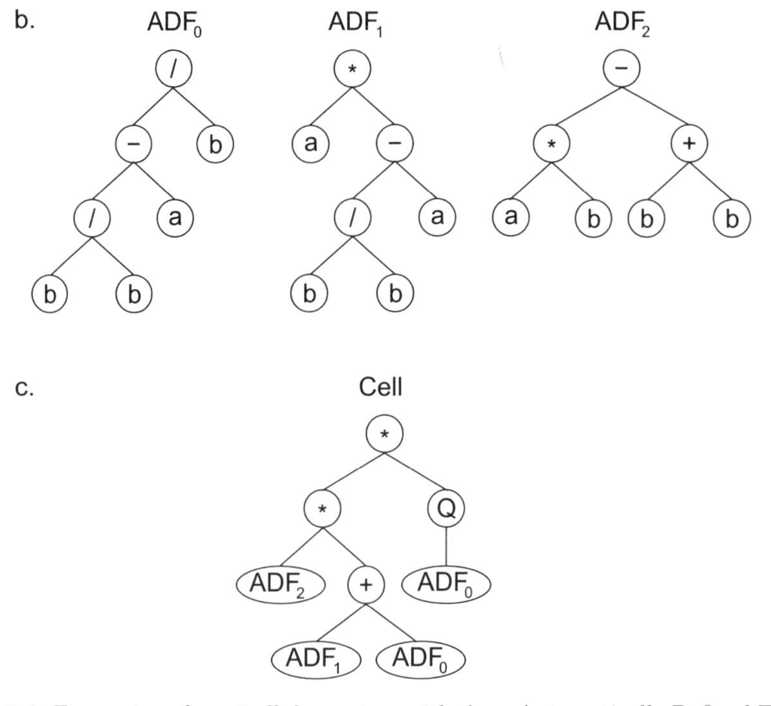

Fig. 2.8. Expression of a unicellular system with three Automatically Defined Functions. **a)** The chromosome composed of three conventional genes and one homeotic gene (shown in bold). **b)** The ADFs codified by each conventional gene. **c)** The main program or cell.

a.
```
012345601234560123456012345678012345678
*Q-bbabQ*baabb-/abbab*+21Q1102/*21+1011
```

b.

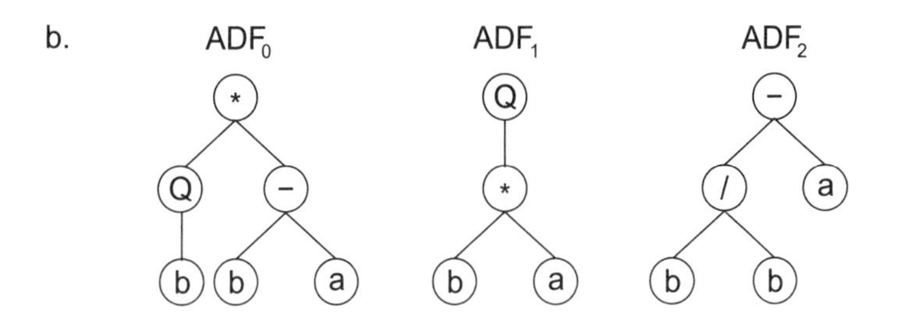

c.

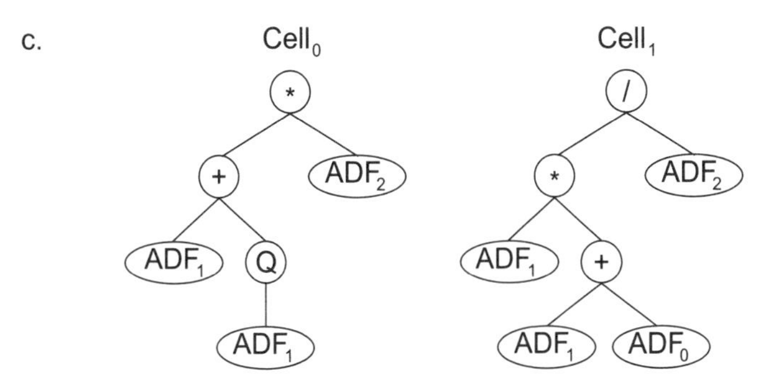

Fig. 2.9. Expression of a multicellular system with three Automatically Defined Functions. **a)** The chromosome composed of three conventional genes and two homeotic genes (shown in bold). **b)** The ADFs codified by each conventional gene. **c)** Two different main programs expressed in two different cells. Note how different cells put together different combinations of ADFs.

It is worth pointing out that the implementation of multiple main programs in Genetic Programming is virtually unthinkable and so far no one has attempted it.

2.4.3 Incorporating Random Numerical Constants in ADFs

The incorporation of random numerical constants in Automatically Defined Functions is also easy and straightforward. As you probably guessed, the gene structure used to accomplish this includes the special domain Dc for encoding the random numerical constants, which, for the sake of simplicity and efficiency, is only implemented in the genes encoding the ADFs (one can obviously extend this organization to the homeotic genes, but nothing is gained from that except a considerable increase in computational effort). The structure of the homeotic genes remains exactly the same and they continue to control how often each ADF is called upon and how they interact with one another.

Consider, for instance, the chromosome with two conventional genes and their respective arrays of random numerical constants:

```
0123456789001234567890012345678012345678
**?b?aa4701+/Q?ba?8536*0Q-10010/Q-+01111
```
(2.13)

$A_0 = \{0.664, 1.703, 1.958, 1.178, 1.258, 2.903, 2.677, 1.761, 0.923, 0.796\}$
$A_1 = \{0.588, 2.114, 0.510, 2.359, 1.355, 0.186, 0.070, 2.620, 2.374, 1.710\}$
The genes encoding the ADFs are expressed exactly as normal genes with a Dc domain and, therefore, their respective ADFs will, most probably, include random numerical constants (Figure 2.10). Then these ADFs with random numerical constants are called upon as many times as necessary from any of the main programs encoded in the homeotic genes. As you can see in Figure 2.10, ADF_0 is invoked twice in Cell_0 and once in Cell_1, whereas ADF_1 is used just once in Cell_0 and called three different times in Cell_1.

2.5 Analyzing the Importance of ADFs in Automatic Programming

The motivation behind the implementation of Automatically Defined Functions in Genetic Programming, was the belief that ADFs allow the evolution of modular solutions and, consequently, improve the performance of the GP technique [13, 14, 15]. Koza proved this by solving a sextic polynomial problem and the even-parity functions, both with and without ADFs [15].

In this section, we are going to solve the sextic polynomial problem using not only a cellular system with ADFs but also a multigenic system with static linking and a simple unigenic system. The study of the simple unigenic system is particularly interesting because it has some similarities with the GP system without ADFs.

2.5.1 General Settings

The sextic polynomial of this section $x^6 - 2x^4 + x^2$ was chosen by Koza [15] because of its potentially exploitable regularity and modularity, easily guessed by its factorization:

$$x^6 - 2x^4 + x^2 = x^2(x-1)^2(x+1)^2 \tag{2.14}$$

For this problem, the basic parameters common to both GEP and GP, irrespective of the presence or absence of ADFs, were kept exactly the same as those used by Koza. Thus, a set of 50 random fitness cases chosen from the interval [-1.0, +1.0] was used; and a very simple function set, composed only

a.
```
01234567890012345678900123456780012345678
**?b?aa4701+/Q?ba?8536*0Q-10010/Q-+01111
```

A_0 = {0.664, 1.703, 1.958, 1.178, 1.258, 2.903, 2.677, 1.761, 0.923, 0.796}
A_1 = {0.588, 2.114, 0.510, 2.359, 1.355, 0.186, 0.070, 2.620, 2.374, 1.710}

b.

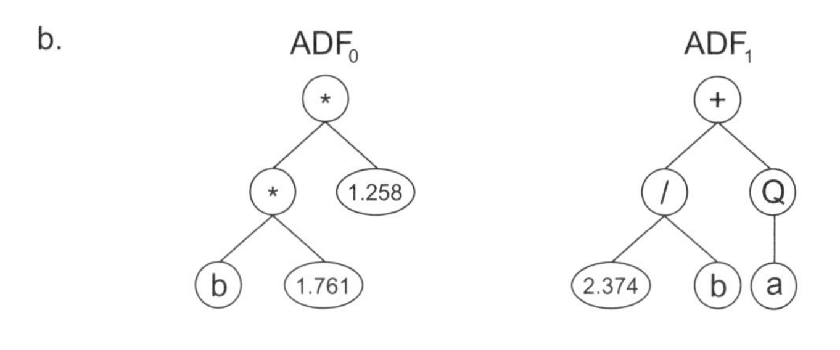

c.

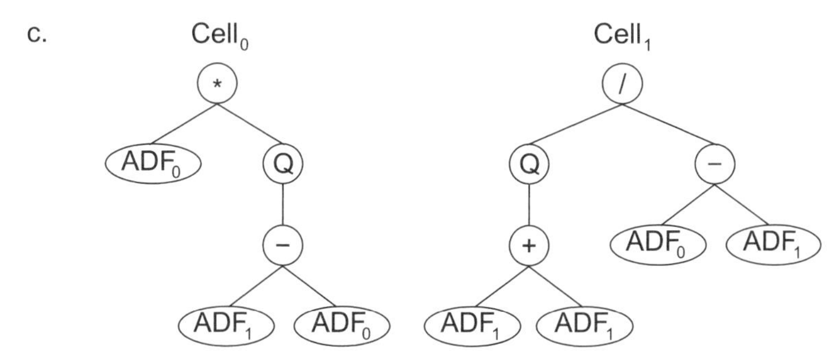

Fig. 2.10. Expression of a multicellular system with Automatically Defined Functions containing random numerical constants. **a)** The chromosome composed of three conventional genes and two homeotic genes (shown in bold). **b)** The ADFs codified by each conventional gene. **c)** Two different programs expressed in two different cells.

of the four arithmetic operators $F = \{+, -, *, /\}$ was used. As for random numerical constants, we will see that their use in this problem is not crucial for the evolution of perfect solutions. Nonetheless, evolution still goes smoothly if integer constants are used and, therefore, one can illustrate the role random constants play and how they are integrated in Automatically Defined Functions by choosing integer random constants. So, when used, integer random constants are chosen from the interval [0, 3].

The fitness function used to evaluate the performance of each evolved program is based on the relative error and explores the idea of a selection range and a precision. The selection range is used as a limit for selection to operate, above which the performance of a program on a particular fitness case contributes nothing to its fitness. And the precision is the limit for improvement, as it allows the fine-tuning of the evolved solutions as accurately as necessary.

Mathematically, the fitness f_i of an individual program i is expressed by the equation:

$$f_i = \sum_{j=1}^{n} \left(R - \left| \frac{P_{(ij)} - T_j}{T_j} \cdot 100 \right| \right) \tag{2.15}$$

where R is the selection range, $P_{(ij)}$ the value predicted by the individual program i for fitness case j (out of n fitness cases) and T_j is the target value for fitness case j. Note that the absolute value term corresponds to the relative error. This term is what is called the precision and if the error is lower than or equal to the precision then the error becomes zero. Thus, for a perfect fit, the absolute value term is zero and $f_i = f_{max} = nR$. In all the experiments of this work we are going to use a selection range of 100 and a precision of 0.01, thus giving for the set of 50 fitness cases $f_{max} = 5{,}000$. It is worth pointing out that these conditions, especially the precision of 0.01, guarantee that all perfect solutions are indeed perfect and match exactly the target function (2.14). This is important to keep in mind since the performance of the different systems will be compared in terms of success rate.

2.5.2 Results without ADFs

The importance of Automatically Defined Functions and the advantages they bring to Evolutionary Computation can only be understood if one analyzes their behavior and how the algorithm copes with their integration. Is evolution still smooth? Are there gains in performance? Is the system still simple or excessively complicated? How does it compare to simpler systems without ADFs? How does one integrate random numerical constants in ADFs? Are these ADFs still manageable or excessively complex? Are there problems that can only be solved with ADFs? These are some of the questions that we will try to address in this work. And for that we are going to start this study by analyzing two simpler systems: the simpler of the two is the unigenic system of GEP that evolves a single parse tree and therefore bares some resemblance to the GP system without ADFs; the other one is the multigenic system of GEP that evolves multiple parse trees linked together by a predefined linking function.

The Unigenic System

For this analysis, we are going to use the basic Gene Expression Algorithm both with and without random constants. In both cases, though, a single tree structure encoded in a single gene will be evolved.

As it is customary, the parameters used per run are summarized in a table (Table 2.1). Note that, in these experiments, small population sizes of only 50 individuals were used, incomparably smaller than the 4,000 individuals used by Koza [15] to solve this same problem (indeed, this population size of 50 individuals is kept constant throughout this work in order to allow a more fruitful comparison between the different systems). Also worth pointing out is that, throughout this study and whenever possible, a maximum program size around 50 points was used so that comparisons could be made between all the experiments (to be more precise, for the unigenic systems a head length of 24 gives maximum program size 49, whereas for the multigenic systems with four genes with head length six, maximum program size equals 52). So, for the unigenic systems of this study, chromosomes with a head length $h = 24$ were chosen, giving maximum program length of 49 (note, however, that the chromosome length in the system with random numerical constants is larger on account of the Dc domain).

As shown in Table 2.1, the probability of success for this problem using the unigenic system without random numerical constants is considerably higher

Table 2.1. Settings for the sextic polynomial problem using a unigenic system with (**ugGEP-RNC**) and without (**ugGEP**) random numerical constants

	ugGEP	ugGEP-RNC
Number of runs	100	100
Number of generations	200	200
Population size	50	50
Chromosome length	49	74
Number of genes	1	1
Head length	24	24
Gene length	49	74
Terminal set	a	a ?
Function set	+ - * /	+ - * /
Mutation rate	0.044	0.044
Inversion rate	0.1	0.1
RIS transposition rate	0.1	0.1
IS transposition rate	0.1	0.1
Two-point recombination rate	0.3	0.3
One-point recombination rate	0.3	0.3
Random constants per gene	–	5
Random constants data type	–	Integer
Random constants range	–	0-3
Dc-specific mutation rate	–	0.044
Dc-specific inversion rate	–	0.1
Dc-specific IS transposition rate	–	0.1
Random constants mutation rate	–	0.01
Number of fitness cases	50	50
Selection range	100	100
Precision	0.01	0.01
Success rate	26%	4%

than the success rate obtained with the ugGEP-RNC algorithm (26% as opposed to just 4%), which, again, shows that, for this kind of simple, modular problem, evolutionary algorithms fare far better if the numerical constants are created from scratch by the algorithm itself through the evolution of simple mathematical expressions. This is not to say that, for complex real-world problems of just one variable or really complex problems of higher dimensions, random numerical constants are not important; indeed, most of the times, they play a crucial role in the discovery of good solutions.

Let's take a look at the structure of the first perfect solution found using the unigenic system without the facility for the manipulation of random numerical constants:

```
0123456789012345678901234567890123456789012345678
*++a/****+-a--*/aa*/*---aaaaaaaaaaaaaaaaaaaaaaaaa     (2.16)
```

As its expression shows, it contains a relatively big neutral region involving a total of nine nodes and two smaller ones involving just three nodes each. It is also worth pointing out the creation of the numerical constant one through the simple arithmetic operation a/a.

It is also interesting to take a look at the structure of the first perfect solution found using the unigenic system with the facility for the manipulation of random numerical constants:

```
0123456789012345678901234567890123456789...
*a+**aa+*a*/aa-a*++-?aa*a?aaa???aaaaaa?a?...

...01234567890123456789012345678901230123     (2.17)
...a?a???a?a0210121021334303442030040
```

```
A = {1, 1, 1, 1, 3}
```

As its expression reveals, it is a fairly compact solution with no obvious neutral regions that makes good use of the random numerical constant one to evolve a perfect solution.

The Multigenic System with Static Linking

For this analysis we are going to use again both the basic Gene Expression Algorithm without random constants and GEP with random numerical constants. The parameters and the performance of both experiments are shown in Table 2.2.

It is worth pointing out that maximum program length in these experiments is similar to the one used in the unigenic systems of the previous section. Here, head lengths $h = 6$ and four genes per chromosome were used, giving maximum program length of 52 points (again note that the chromosome length in the systems with random numerical constants is larger on account of the Dc domain, but maximum program length remains the same).

Table 2.2. Settings for the sextic polynomial problem using a multigenic system with (**mgGEP-RNC**) and without (**mgGEP**) random numerical constants

	mgGEP	mgGEP-RNC
Number of runs	100	100
Number of generations	200	200
Population size	50	50
Chromosome length	52	80
Number of genes	4	4
Head length	6	6
Gene length	13	20
Linking function	*	*
Terminal set	a	a ?
Function set	+ - * /	+ - * /
Mutation rate	0.044	0.044
Inversion rate	0.1	0.1
RIS transposition rate	0.1	0.1
IS transposition rate	0.1	0.1
Two-point recombination rate	0.3	0.3
One-point recombination rate	0.3	0.3
Gene recombination rate	0.3	0.3
Gene transposition rate	0.1	0.1
Random constants per gene	–	5
Random constants data type	–	Integer
Random constants range	–	0-3
Dc-specific mutation rate	–	0.044
Dc-specific inversion rate	–	0.1
Dc-specific IS transposition rate	–	0.1
Random constants mutation rate	–	0.01
Number of fitness cases	50	50
Selection range	100	100
Precision	0.01	0.01
Success rate	93%	49%

As you can see by comparing Tables 2.1 and 2.2, the use of multiple genes resulted in a considerable increase in performance for both systems. In the systems without random numerical constants, by partitioning the genome into four autonomous genes, the performance increased from 26% to 93%, whereas in the systems with random numerical constants, the performance increased from 4% to 49%. Note also that, in this analysis, the already familiar pattern is observed when random numerical constants are introduced: the success rate decreases considerably from 93% to 49% (in the unigenic systems it decreased from 26% to 4%).

Let's also take a look at the structure of the first perfect solution found using the multigenic system without the facility for the manipulation of random numerical constants (the sub-ETs are linked by multiplication):

```
0123456789012012345678901201234567890120123456789012
+/aaa/aaaaaaa+//a/aaaaaaaa-**a/aaaaaaaa-**a/aaaaaaaa
```
(2.18)

As its expression shows, it contains three small neutral regions involving a total of nine nodes, all encoding the numerical constant one. Note also that, in two occasions (in sub-ETs 0 and 1), the numerical constant one plays an important role in the overall making of the perfect solution. Also interesting about this perfect solution, is that genes 2 and 3 are exactly the same, suggesting a major event of gene duplication (it's worth pointing out that the duplication of genes can only be achieved by the concerting action of gene recombination and gene transposition, as a gene duplication operator is not part of the genetic modification arsenal of Gene Expression Programming).

It is also interesting to take a look at the structure of the first perfect solution found using the multigenic system with the facility for the manipulation of random numerical constants (the sub-ETs are linked by multiplication):

```
01234567890123456789
+--+*aa??aa?a0444212
+--+*aa??aa?a0244422
a?a??a?aaa?a?2212021
aa-a*/?aa????3202123

A0 = {0, 3, 1, 2, 1}
A1 = {0, 3, 1, 2, 1}
A2 = {0, 3, 1, 2, 1}
A3 = {3, 3, 2, 0, 2}
```
(2.19)

As its expression reveals, it is a fairly compact solution with two small neutral motifs plus a couple of neutral nodes, all representing the numerical constant zero. Note that genes 0 and 1 are almost exact copies of one another (there is only variation at positions 17 and 18, but they are of no consequence as they are part of a noncoding region of the gene), suggesting a recent event of gene duplication. Note also that although genes 2 and 3 encode exactly the same sub-ET (a simple sub-ET with just one node), they most certainly followed different evolutionary paths as the homology between their sequences suggests.

2.5.3 Results with ADFs

In this section, we are going to conduct a series of four studies, each with four different experiments. In the first study, we are going to use a unicellular system encoding 1, 2, 3, and 4 ADFs. In the second study, we are going to use again a unicellular system, encoding also 1, 2, 3, and 4 ADFs, but, in this case, the ADFs will also incorporate random numerical constants. The third and fourth studies are respectively similar to the first and second one, with the difference that we are going to use a system with multiple cells (three, to be precise) instead of just the one.

The Unicellular System

For the unicellular system without random numerical constants, both the parameters and performance are shown in Table 2.3. It is worth pointing out, that, in these experiments, we are dealing with Automatically Defined Functions that can be reused again and again, and therefore it makes little sense to talk about maximum program length. However, in these series of experiments the same head length of six was used to encode all the ADFs and a system with four ADFs was also analyzed, thus allowing the comparison of these cellular systems with the simpler acellular ones (recall that we used four genes with $h = 6$ in the multigenic system and one gene with $h = 24$ in the unigenic system).

As you can see in Table 2.3, these cellular systems with Automatically Defined Functions perform extremely well, especially if we compare them with the unigenic system (see Table 2.1), which, as you recall, is the closer we can get to the Genetic Programming system. So, we can also conclude that the use of ADFs can bring considerable benefits to systems limited to just one gene or parse tree, especially if there is some modularity in the problem at hand. Note, however, that the unicellular system slightly pales in comparison to the multigenic system with static linking (see Table 2.2), showing that the multigenic system of GEP is already a highly efficient system that can be used to solve virtually any kind of problem.

It is worth pointing out that, in his analysis of the role of ADFs to solve this problem, Koza [15] uses just one ADF, which, as you can see in Table 2.3, is the most successful organization to solve this problem, with a success rate of 82%. And, curiously enough, this same pattern appears in all the experiments conducted here, in which the highest success rates were obtained when just one ADF was used (see Tables 2.3, 2.4, 2.5, and 2.6).

Let's take a look at the structure of the first perfect solution found using the unicellular system encoding just one ADF:

```
0123456789012012345678
-a**aaaaaaaaa*/0*00000
```
(2.20)

As its expression reveals, the main program is far from parsimonious and could be simplified to $(\text{ADF})^2$. But, nevertheless, it illustrates perfectly how a useful building block created by the evolutionary process itself can be reused several times by the main program encoded in the homeotic gene.

Let's also analyze the structure of a program with more than one ADF, the individual below with four ADFs (genes are shown separately):

```
0123456789012
*-a--+aaaaaaa
/-a---aaaaaaa
-a*+*-aaaaaaa
a+*-+aaaaaaaa
*22121133
```
(2.21)

As its expression shows, the main program is fairly compact and invokes only ADF_2. All the remaining ADFs are not used at all, and, therefore, are free to evolve without much pressure. We know already that neutral regions play an important role both in natural evolution [12] and in GEP [5], and that their use in good measure is responsible for a considerable increase in performance. And the same phenomenon is observed in these cellular systems, where the simplest one with only one ADF and a single cell seems to have the right amount of neutral regions as it evolves very efficiently with a success rate of 82%. So, in this particular problem, by increasing the number of ADFs, we are basically increasing the number of neutral regions, and the performance

Table 2.3. Settings and performance for the sextic polynomial problem using a unicellular system encoding 1, 2, 3, and 4 ADFs

	1 ADF	2 ADFs	3 ADFs	4 ADFs
Number of runs	100	100	100	100
Number of generations	200	200	200	200
Population size	50	50	50	50
Chromosome length	22	35	48	61
Number of genes/ADFs	1	2	3	4
Head length	6	6	6	6
Gene length	13	13	13	13
Function set of ADFs	+ - * /	+ - * /	+ - * /	+ - * /
Terminal set	a	a	a	a
Number of homeotic genes/cells	1	1	1	1
Head length of homeotic genes	4	4	4	4
Length of homeotic genes	9	9	9	9
Function set of homeotic genes	+ - * /	+ - * /	+ - * /	+ - * /
Terminal set of homeotic genes	ADF 0	ADFs 0-1	ADFs 0-2	ADFs 0-3
Mutation rate	0.044	0.044	0.044	0.044
Inversion rate	0.1	0.1	0.1	0.1
RIS transposition rate	0.1	0.1	0.1	0.1
IS transposition rate	0.1	0.1	0.1	0.1
Two-point recombination rate	0.3	0.3	0.3	0.3
One-point recombination rate	0.3	0.3	0.3	0.3
Gene recombination rate	0.3	0.3	0.3	0.3
Gene transposition rate	–	0.1	0.1	0.1
Mutation rate in homeotic genes	0.044	0.044	0.044	0.044
Inversion rate in homeotic genes	0.1	0.1	0.1	0.1
RIS transp. in homeotic genes	0.1	0.1	0.1	0.1
IS transp. in homeotic genes	0.1	0.1	0.1	0.1
Number of fitness cases	50	50	50	50
Selection range	100	100	100	100
Precision	0.01	0.01	0.01	0.01
Success rate	82%	78%	69%	63%

for this simple modular problem decreases proportionately, dropping down to 63% in the system with four ADFs (see Table 2.3).

Let's now analyze the behavior of the unicellular system when random numerical constants are also incorporated in the Automatically Defined Functions.

For that purpose a similar set of experiments were done, using also 1, 2, 3, and 4 ADFs (Table 2.4). And as expected, a considerable decrease in performance was observed comparatively to the performance observed in the unicellular system without random numerical constants (see Table 2.3).

Let's take a look at the structure of the first perfect solution found using the unicellular system encoding just one ADF with random constants:

```
01234567890123456789012345678
---*a-?aa?a??0412201*+0*00000
                                  (2.22)
A = {1, 0, 3, 1, 2}
```

As its expression shows, the simple module discovered in the structure encoding the ADF (a^2-1) is called four times in the main program, creating a perfect solution with just one kind of building block.

Let's now analyze the structure of a program with more than one ADF, the individual below with four ADFs with random numerical constants (the genes are shown separately):

```
01234567890123456789
*-+*a*?aaa?a?3324010
-aa?*-aa?a???3123440
*a/a-a?aa?a??2234201
--*+*+??aa?aa0141122
*00003233                         (2.23)

A0 = {1, 1, 2, 1, 1}
A1 = {2, 3, 0, 2, 0}
A2 = {3, 1, 0, 0, 3}
A3 = {2, 0, 3, 2, 2}
```

As its expression reveals, the main program is fairly compact and invokes only ADF_0. Indeed, for this simple modular problem, most perfect solutions involve just one ADF, suggesting that this problem is better solved using just one kind of building block that can then be used as many times as necessary by the main program. And the fact that the system evolves more efficiently with just one ADF is just another indication of this (57% success rate in the system with just one ADF versus 26% in the system with four ADFs).

Table 2.4. Settings and performance for the sextic polynomial problem using a unicellular system encoding 1, 2, 3, and 4 ADFs with random numerical constants

	1 ADF	2 ADFs	3 ADFs	4 ADFs
Number of runs	100	100	100	100
Number of generations	200	200	200	200
Population size	50	50	50	50
Chromosome length	29	49	69	89
Number of genes/ADFs	1	2	3	4
Head length	6	6	6	6
Gene length	20	20	20	20
Function set of ADFs	+ - * /	+ - * /	+ - * /	+ - * /
Terminal set	a ?	a ?	a ?	a ?
Number of homeotic genes/cells	1	1	1	1
Head length of homeotic genes	4	4	4	4
Length of homeotic genes	9	9	9	9
Function set of homeotic genes	+ - * /	+ - * /	+ - * /	+ - * /
Terminal set of homeotic genes	ADF 0	ADFs 0-1	ADFs 0-2	ADFs 0-3
Mutation rate	0.044	0.044	0.044	0.044
Inversion rate	0.1	0.1	0.1	0.1
RIS transposition rate	0.1	0.1	0.1	0.1
IS transposition rate	0.1	0.1	0.1	0.1
Two-point recombination rate	0.3	0.3	0.3	0.3
One-point recombination rate	0.3	0.3	0.3	0.3
Gene recombination rate	0.3	0.3	0.3	0.3
Gene transposition rate	–	0.1	0.1	0.1
Random constants per gene	5	5	5	5
Random constants data type	Integer	Integer	Integer	Integer
Random constants range	0-3	0-3	0-3	0-3
Dc-specific mutation rate	0.044	0.044	0.044	0.044
Dc-specific inversion rate	0.1	0.1	0.1	0.1
Dc-specific IS transposition rate	0.1	0.1	0.1	0.1
Random constants mutation rate	0.01	0.01	0.01	0.01
Mutation rate in homeotic genes	0.044	0.044	0.044	0.044
Inversion rate in homeotic genes	0.1	0.1	0.1	0.1
RIS transp. in homeotic genes	0.1	0.1	0.1	0.1
IS transp. in homeotic genes	0.1	0.1	0.1	0.1
Number of fitness cases	50	50	50	50
Selection range	100	100	100	100
Precision	0.01	0.01	0.01	0.01
Success rate	57%	42%	33%	26%

The Multicellular System

For the multicellular system without random numerical constants, both the parameters and performance are shown in Table 2.5.

As you can see, these multicellular systems with Automatically Defined Functions perform extremely well, even better than the multigenic system with static linking (see Table 2.2). And they are very interesting because they can also be used to solve problems with multiple outputs, where each cell is engaged in the identification of one class or output. Here, however, we are using a multicellular system to solve a problem with just one output, which means that all the cells are trying to find the same kind of solution and, therefore, for each individual, the fitness is determined by the best cell. Obviously, the greater the number of cells the higher the probability of evolving the perfect solution or cell. But there is one caveat though: one cannot go on increasing

Table 2.5. Settings and performance for the sextic polynomial problem using a multicellular system encoding 1, 2, 3, and 4 ADFs

	1 ADF	2 ADFs	3 ADFs	4 ADFs
Number of runs	100	100	100	100
Number of generations	200	200	200	200
Population size	50	50	50	50
Chromosome length	40	53	66	79
Number of genes/ADFs	1	2	3	4
Head length	6	6	6	6
Gene length	13	13	13	13
Function set of ADFs	+ - * /	+ - * /	+ - * /	+ - * /
Terminal set	a	a	a	a
Number of homeotic genes/cells	3	3	3	3
Head length of homeotic genes	4	4	4	4
Length of homeotic genes	9	9	9	9
Function set of homeotic genes	+ - * /	+ - * /	+ - * /	+ - * /
Terminal set of homeotic genes	ADF 0	ADFs 0-1	ADFs 0-2	ADFs 0-3
Mutation rate	0.044	0.044	0.044	0.044
Inversion rate	0.1	0.1	0.1	0.1
RIS transposition rate	0.1	0.1	0.1	0.1
IS transposition rate	0.1	0.1	0.1	0.1
Two-point recombination rate	0.3	0.3	0.3	0.3
One-point recombination rate	0.3	0.3	0.3	0.3
Gene recombination rate	0.3	0.3	0.3	0.3
Gene transposition rate	–	0.1	0.1	0.1
Mutation rate in homeotic genes	0.044	0.044	0.044	0.044
Inversion rate in homeotic genes	0.1	0.1	0.1	0.1
RIS transp. in homeotic genes	0.1	0.1	0.1	0.1
IS transp. in homeotic genes	0.1	0.1	0.1	0.1
Number of fitness cases	50	50	50	50
Selection range	100	100	100	100
Precision	0.01	0.01	0.01	0.01
Success rate	98%	96%	95%	90%

the number of cells indefinitely because it takes time and resources to express all of them. The use of three cells per individual seems a good compromise, and we are going to use just that in all the experiments of this section.

Let's take a look at the structure of the first perfect solution found using the multicellular system encoding just one ADF (which cell is best is indicated after the coma):

```
0123456789012012345678012345678012345678
-*a*aaaaaaaaa-0-+000000***00000-**+00000,2
```
(2.24)

As its expression shows, the best main program invokes the ADF encoded in the conventional gene five times. Note, however, that this perfect solution is far from parsimonious and could indeed be simplified to $(\text{ADF})^2$.

It is also interesting to take a look at what the other cells are doing. For instance, Cell_0 encodes zero and Cell_1 encodes $(\text{ADF})^4$, both a far cry from the perfect solution.

Let's also analyze the structure of a program with more than one ADF, the individual below with four ADFs and three cells (the best cell is indicated after the coma):

```
0123456789012012345678901201234567890120123456789012
**-/aaaaaaaaa/aaaa/aaaaaaa*-a*/aaaaaaaa/-***/aaaaaaa

012345678012345678012345678
*-*222301*+2021323-+0020321,1
```
(2.25)

As its expression shows, the best main program invokes two different ADFs (ADF_0 and ADF_2), but since ADF_0 encodes zero, the best cell could be simplified to $(\text{ADF}_2)^2$, which is again a perfect solution built with just one kind of building block ($a^3 - a$). It is also worth noticing that two of the ADFs (ADF_0 and ADF_3) and one of the cells (Cell_0) encode zero, and the numerical constant one is also encoded by ADF_1; they are all good examples of the kind of neutral region that permeates all these solutions.

Let's now analyze the behavior of the multicellular system when random numerical constants are also incorporated in the Automatically Defined Functions.

For that purpose a similar set of experiments were done, using also 1, 2, 3, and 4 ADFs (Table 2.6). And as expected, a considerable decrease in performance was observed comparatively to the performance obtained in the multicellular system without random numerical constants (see Table 2.5). Notwithstanding, ADFs with random numerical constants perform quite well despite the additional complexity, and they may prove valuable in problems where random numerical constants are crucial to the discovery of good solutions.

Table 2.6. Settings and performance for the sextic polynomial problem using a multicellular system encoding 1, 2, 3, and 4 ADFs with random numerical constants

	1 ADF	2 ADFs	3 ADFs	4 ADFs
Number of runs	100	100	100	100
Number of generations	200	200	200	200
Population size	50	50	50	50
Chromosome length	29	67	87	107
Number of genes/ADFs	1	2	3	4
Head length	6	6	6	6
Gene length	20	20	20	20
Function set of ADFs	+ - * /	+ - * /	+ - * /	+ - * /
Terminal set	a ?	a ?	a ?	a ?
Number of homeotic genes/cells	1	3	3	3
Head length of homeotic genes	4	4	4	4
Length of homeotic genes	9	9	9	9
Function set of homeotic genes	+ - * /	+ - * /	+ - * /	+ - * /
Terminal set of homeotic genes	ADF 0	ADFs 0-1	ADFs 0-2	ADFs 0-3
Mutation rate	0.044	0.044	0.044	0.044
Inversion rate	0.1	0.1	0.1	0.1
RIS transposition rate	0.1	0.1	0.1	0.1
IS transposition rate	0.1	0.1	0.1	0.1
Two-point recombination rate	0.3	0.3	0.3	0.3
One-point recombination rate	0.3	0.3	0.3	0.3
Gene recombination rate	0.3	0.3	0.3	0.3
Gene transposition rate	–	0.1	0.1	0.1
Random constants per gene	5	5	5	5
Random constants data type	Integer	Integer	Integer	Integer
Random constants range	0-3	0-3	0-3	0-3
Dc-specific mutation rate	0.044	0.044	0.044	0.044
Dc-specific inversion rate	0.1	0.1	0.1	0.1
Dc-specific IS transposition rate	0.1	0.1	0.1	0.1
Random constants mutation rate	0.01	0.01	0.01	0.01
Mutation rate in homeotic genes	0.044	0.044	0.044	0.044
Inversion rate in homeotic genes	0.1	0.1	0.1	0.1
RIS transp. in homeotic genes	0.1	0.1	0.1	0.1
IS transp. in homeotic genes	0.1	0.1	0.1	0.1
Number of fitness cases	50	50	50	50
Selection range	100	100	100	100
Precision	0.01	0.01	0.01	0.01
Success rate	79%	60%	58%	50%

Let's take a look at the structure of a perfect solution found using the multicellular system encoding just one ADF with random numerical constants (the best cell is indicated after the coma):

```
01234567890123456789012345678012345678012345678
-***a?aaa???a4000424+/0+00000*0*/00000-+*-00000,1
```
(2.26)

```
A = {3, 1, 1, 1, 1}
```

As its expression shows, the main program encoded in $Cell_1$ is far from parsimonious, but it encodes nonetheless a perfect solution to the sextic polynomial (2.14). The only available ADF is called four times from the main program, but in essence it could have been called just twice as it can be simplified to $(ADF)^2$.

Let's now analyze the structure of a program with more than one ADF, the individual below with four ADFs and three cells (the best cell is indicated after the coma):

```
01234567890123456789
-a*-*a?aa????2322013
?-*//aaa?aa?a2412442
*a+a*aa?aaaaa4024010
*-a?a?aaa????3224232

A0 = {0, 0, 0, 1, 0}
A1 = {2, 0, 0, 2, 2}
A2 = {2, 1, 3, 0, 0}
A3 = {2, 1, 3, 0, 0}

012345678012345678012345678
*2*/00213/-+*03022233201102,0
```
(2.27)

As its expression shows, the best main program invokes two different ADFs (ADF_0 and ADF_2), but the calls to ADF_2 cancel themselves out, and the main program is reduced to $(ADF_0)^2$, which, of course is a perfect solution to the problem at hand.

2.6 Summary

Comparatively to Genetic Programming, the implementation of Automatically Defined Functions in Gene Expression Programming is very simple because it stands on the shoulders of the multigenic system with static linking and, therefore, requires just a small addition to make it work. And because the cellular system of GEP with ADFs, like all GEP systems, continues to be totally encoded in a simple linear genome, it poses no constraints whatsoever to the action of the genetic operators and, therefore, these systems can also evolve efficiently (indeed, all the genetic operators of GEP were easily extended to the homeotic genes). As a comparison, the implementation of ADFs in GP adds additional constraints to the already constrained genetic operators in order to ensure the integrity of the different structural branches of the parse tree. Furthermore, due to its mammothness, the implementation of multiple main programs in Genetic Programming is prohibitive, whereas in GEP the creation of a multicellular system encoding multiple main programs is a child's play.

Indeed, another advantage of the cellular system of GEP, is that it can easily grow into a multicellular one, encoding not just one but multiple cells or main programs, each using a different set of ADFs. These multicellular systems have multiple applications, some of which were already illustrated in this work, but their real potential resides in solving problems with multiple outputs where each cell encodes a program involved in the identification of a certain class or pattern. Indeed, the high performance exhibited by the multicellular system in this work gives hope that this system can be fruitfully explored to solve much more complex problems. In fact, in this work, not only the multicellular but also the unicellular and the multigenic system with static linking, were all far from stretched to their limits as the small population sizes of just 50 individuals used in all the experiments of this work indicate. As a comparison, to solve this same problem, the GP system with ADFs uses already populations of 4,000 individuals.

And yet another advantage of the ADFs of Gene Expression Programming, is that they are free to become functions of one or several arguments, being this totally decided by evolution itself. Again, in GP, the number of arguments each ADF takes must be a priori decided and cannot be changed during the course of evolution lest invalid structures are created.

And finally, the cellular system (and multicellular also) encoding ADFs with random numerical constants was for the first time described in this work. Although their performance was also compared to other systems, the main goal was to show that ADFs with random numerical constants can also evolve efficiently, extending not only their appeal but also the range of their potential applications.

References

1. N.L. Cramer, A Representation for the Adaptive Generation of Simple Sequential Programs. In J. J. Grefenstette, ed., "Proceedings of the First International Conference on Genetic Algorithms and Their Applications", Erlbaum, 1985
2. R. Dawkins, River out of Eden, Weidenfeld and Nicolson, 1995
3. C. Ferreira, Gene Expression Programming: A New Adaptive Algorithm for Solving Problems, Complex Systems, 13 (2): 87-129, 2001
4. C. Ferreira, Gene Expression Programming: Mathematical Modeling by an Artificial Intelligence, Angra do Heroısmo, Portugal, 2002
5. C. Ferreira, Genetic Representation and Genetic Neutrality in Gene Expression Programming, Advances in Complex Systems, 5 (4): 389-408, 2002
6. C. Ferreira, Mutation, Transposition, and Recombination: An Analysis of the Evolutionary Dynamics. In H.J. Caulfield, S.-H. Chen, H.-D. Cheng, R. Duro, V. Honavar, E.E. Kerre, M. Lu, M. G. Romay, T. K. Shih, D. Ventura, P.P.

Wang, Y. Yang, eds., "Proceedings of the 6th Joint Conference on Information Sciences, 4th International Workshop on Frontiers in Evolutionary Algorithms", pp. 614-617, Research Triangle Park, North Carolina, USA, 2002
7. C. Ferreira, Gene Expression Programming and the Evolution of Computer Programs. In L.N. de Castro and F.J. Von Zuben, eds., "Recent Developments in Biologically Inspired Computing", pp. 82-103, Idea Group Publishing, 2004
8. C. Ferreira, Designing Neural Networks Using Gene Expression Programming, 9th Online World Conference on Soft Computing in Industrial Applications, September 20 - October 8, 2004
9. R.M. Friedberg, A Learning Machine: Part I, IBM Journal, 2 (1): 2-13, 1958
10. R.M. Friedberg, B. Dunham, and J.H. North, A Learning Machine: Part II, IBM Journal, 3 (7): 282-287, 1959
11. J.H. Holland, Adaptation in Natural and Artificial Systems: An Introductory Analysis with Applications to Biology, Control, and Artificial Intelligence, University of Michigan Press, USA, 1975 (second edition: MIT Press, 1992)
12. M. Kimura, The Neutral Theory of Molecular Evolution, Cambridge University Press, Cambridge, UK, 1983
13. J.R. Koza, F.H. Bennett III, D. Andre, and M.A. Keane, Genetic Programming III: Darwinian Invention and Problem Solving, Morgan Kaufmann Publishers, San Francisco, USA, 1999
14. J.R. Koza, Genetic Programming: On the Programming of Computers by Means of Natural Selection, MIT Press, Cambridge, MA, USA, 1992
15. J.R. Koza, Genetic Programming II: Automatic Discovery of Reusable Programs, MIT Press, Cambridge, MA, USA, 1994

3

Evolving Intrusion Detection Systems

Ajith Abraham[1] and Crina Grosan[2]

[1] School of Computer Science and Engineering, Chung-Ang University, 221, Heukseok-Dong, Dongjak-Gu, Seoul, 156-756, Korea
ajith.abraham@ieee.org, http://ajith.softcomputing.net

[2] Department of Computer Science, Faculty of Mathematics and Computer Science, Babeş Bolyai University, Kogalniceanu 1, Cluj-Napoca, 3400, Romania
cgrosan@cs.ubbcluj.ro, http://www.cs.ubbcluj.ro/ cgrosan

An intrusion Detection System (IDS) is a program that analyzes what happens or has happened during an execution and tries to find indications that the computer has been misused. An IDS does not eliminate the use of preventive mechanism but it works as the last defensive mechanism in securing the system. This Chapter evaluates the performances of two Genetic Programming techniques for IDS namely Linear Genetic Programming (LGP) and Multi-Expression Programming (MEP). Results are then compared with some machine learning techniques like Support Vector Machines (SVM) and Decision Trees (DT). Empirical results clearly show that GP techniques could play an important role in designing real time intrusion detection systems.

3.1 Introduction

Computer security is defined as the protection of computing systems against threats to confidentiality, integrity, and availability [28]. Confidentiality (or secrecy) means that information is disclosed only according to policy, integrity means that information is not destroyed or corrupted and that the system performs correctly, availability means that system services are available when they are needed. Computing system refers to computers, computer networks, and the information they handle. Security threats come from different sources such as natural forces (such as flood), accidents (such as fire), failure of services (such as power) and people known as intruders. There are two types of intruders: the external intruders who are unauthorized users of the machines they attack, and internal intruders, who have permission to access the system with some restrictions. The traditional prevention techniques such as user authentication, data encryption, avoiding programming errors and firewalls are

A. Abraham and C. Grosan: *Evolving Intrusion Detection Systems*, Studies in Computational Intelligence (SCI) **13**, 57–79 (2006)
www.springerlink.com

used as the first line of defense for computer security. If a password is weak and is compromised, user authentication cannot prevent unauthorized use, firewalls are vulnerable to errors in configuration and ambiguous or undefined security policies. They are generally unable to protect against malicious mobile code, insider attacks and unsecured modems. Programming errors cannot be avoided as the complexity of the system and application software is changing rapidly leaving behind some exploitable weaknesses. intrusion detection is therefore required as an additional wall for protecting systems. Intrusion detection is useful not only in detecting successful intrusions, but also provides important information for timely countermeasures.

An intrusion is defined [10] as any set of actions that attempt to compromise the integrity, confidentiality or availability of a resource. This includes a deliberate unauthorized attempt to access information, manipulate information, or render a system unreliable or unusable. An attacker can gain illegal access to a system by fooling an authorized user into providing information that can be used to break into a system. An attacker can deliver a piece of software to a user of a system which is actually a trojan horse containing malicious code that gives the attacker system access. Bugs in trusted programs can be exploited by an attacker to gain unauthorized access to a computer system. There are legitimate actions that one can perform that when taken to the extreme can lead to system failure. An attacker can gain access because of an error in the configuration of a system. In some cases it is possible to fool a system into giving access by misrepresenting oneself. An example is sending a TCP packet that has a forged source address that makes the packet appear to come from a trusted host. Intrusions are classified [29] as six types.

Attempted break-ins, which are detected by typical behavior profiles or violations of security constraints. Masquerade attacks, which are detected by atypical behavior profiles or violations of security constraints. Penetration of the security control system, which are detected by monitoring for specific patterns of activity. Leakage, which is detected by atypical use of system resources. Denial of service, which is detected by a typical use of system resources. Malicious use, which is detected by atypical behavior profiles, violations of security constraints, or use of special privileges.

3.2 Intrusion Detection

Intrusion detection is classified into two types: misuse intrusion detection and anomaly intrusion detection. Misuse intrusion detection uses well-defined patterns of the attack that exploit weaknesses in system and application software to identify the intrusions. These patterns are encoded in advance and used to match against the user behavior to detect intrusion.

Anomaly intrusion detection uses the normal usage behavior patterns to identify the intrusion. The normal usage patterns are constructed from the statistical measures of the system features, for example, the CPU and I/O activities by a particular user or program. The behavior of the user is observed and any deviation from the constructed normal behavior is detected as intrusion.

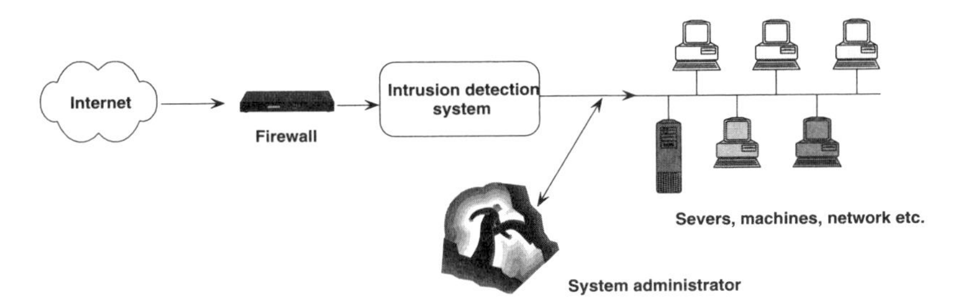

Fig. 3.1. Network protected by an IDS

Figure 3.1 illustrates a simple network, which is protected using IDS. We have two options to secure the system completely, either prevent the threats and vulnerabilities which come from flaws in the operating system as well as in the application programs or detect them and take some action to prevent them in future and also repair the damage. It is impossible in practice, and even if possible, extremely difficult and expensive, to write a completely secure system. Transition to such a system for use in the entire world would be an equally difficult task. Cryptographic methods can be compromised if the passwords and keys are stolen. No matter how secure a system is, it is vulnerable to insiders who abuse their privileges. There is an inverse relationship between the level of access control and efficiency. More access controls make a system less user-friendly and more likely of not being used. An Intrusion Detection system is a program (or set of programs) that analyzes what happens or has happened during an execution and tries to find indications that the computer has been misused. An Intrusion detection system does not eliminate the use of preventive mechanism but it works as the last defensive mechanism in securing the system.

Data mining approaches are a relatively new techniques for intrusion detection. There are a wide variety of data mining algorithms drawn from the fields of statistics, pattern recognition, machine learning, and databases. Previous research of data mining approaches for intrusion detection model identified several types of algorithms as useful techniques. Classification is one of the data mining algorithms, which have been investigated as a useful technique for intrusion detection models.

3.3 Related Research

James Anderson [2] first proposed that audit trails should be used to monitor threats. All the available system security procedures were focused on denying access to sensitive data from an unauthorized source. Dorothy Denning [7] first proposed the concept of intrusion detection as a solution to the problem of providing a sense of security in computer systems. The basic idea is that intrusion behavior involves abnormal usage of the system. The model is a rule-based pattern matching system. Some models of normal usage of the system could be constructed and verified against usage of the system and any significant deviation from the normal usage flagged as abnormal usage. This model served as an abstract model for further developments in this field and is known as generic intrusion detection model and is depicted in Figure 3.2 [15].

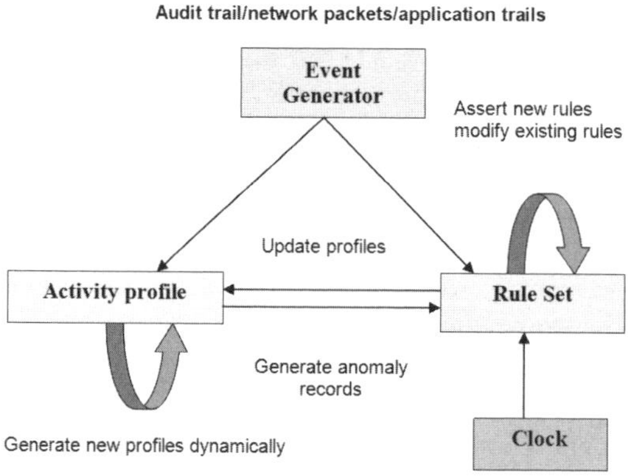

Fig. 3.2. A generic intrusion detection model

Statistical approaches compare the recent behavior of a user of a computer system with observed behavior and any significant deviation is considered as intrusion. This approach requires construction of a model for normal user behavior. IDES (Intrusion Detection Expert System) [17] first exploited the statistical approach for the detection of intruders. It uses the intrusion detection model proposed by Denning [7] and audit trails data as suggested in Anderson [2]. IDES maintains profiles, which is a description of a subject's normal behavior with respect to a set of intrusion detection measures. Profiles are updated periodically, thus allowing the system to learn new behavior as users alter their behavior. These profiles are used to compare the user behavior and informing significant deviation from them as the intrusion. IDES also uses the expert system concept to detect misuse intrusions. This system has later developed as NIDES (Next-generation Intrusion Detection Expert System) [18]. The advantage of this approach is that it adaptively learns the behavior of users, which is thus potentially more sensitive than human experts. This system has several disadvantages. The system can be trained for certain behavior gradually making the abnormal behavior as normal, which makes intruders undetected. Determining the threshold above which an intrusion should be detected is a difficult task. Setting the threshold too low results in false positives (normal behavior detected as an intrusion) and setting too high results in false negatives (an intrusion undetected). Attacks, which occur by sequential dependencies, cannot be detected, as statistical analysis is insensitive to order of events.

Predictive pattern generation uses a rule base of user profiles defined as statistically weighted event sequences [30]. This method of intrusion detection attempts to predict future events based on events that have already occurred. This system develops sequential rules of the from E1 - E2 - E3 $\rightarrow$ (E4 = 94%; E5 = 6%) where the various E's are events derived from the security audit trail, and the percentage on the right hand of the rule represent the probability of occurrence of each of the consequent events given the occurrence of the antecedent sequence. This would mean that for the sequence of observed events E1 followed by E2 followed by E3, the probability of event E4 occurring is 94% and that of E5 is 6%. The rules are generated inductively with an information theoretic algorithm that measures the applicability of rules in terms of coverage and predictive power. An intrusion is detected if the observed sequence of events matches the left hand side of the rule but the following events significantly deviate from the right hand side of the rule. The main advantages of this approach include its ability to detect and respond quickly to anomalous behavior, easier to detect users who try to train the system during its learning period. The main problem with the system is its inability to detect some intrusions if that particular sequence of events have not been recognized and created into the rules.

State transition analysis approach construct the graphical representation of intrusion behavior as a series of state changes that lead from an initial secure state to a target compromised state. Using the audit trail as input, an analysis tool can be developed to compare the state changes produced by the user to state transition diagrams of known penetrations. State transition diagrams form the basis of a rule-based expert system for detecting penetrations, called the State Transition Analysis Tool (STAT) [23]. The STAT prototype is implemented in USTAT (Unix State Transition Analysis Tool) [11] on UNIX based system. The main advantage of the method is it detects the intrusions independent of audit trial record. The rules are produced from the effects of sequence of audit trails on system state whereas in rule based methods the sequence of audit trails are used. It is also able to detect cooperative attacks, variations to the known attacks and attacks spanned across multiple user sessions. Disadvantages of the system are it can only construct patterns from sequence of events but not from more complex forms and some attacks cannot be detected, as they cannot be modeled with state transitions.

Keystroke monitoring technique utilizes a user's keystrokes to determine the intrusion attempt. The main approach is to pattern match the sequence of keystrokes to some predefined sequences to detect the intrusion. The main problems with this approach are lack of support from operating system to capture the keystroke sequences and also many ways of expressing the sequence of keystrokes for the same attack. Some shell programs like *bash*, *ksh* have the user definable aliases utility. These aliases make this technique difficult to detect the intrusion attempts unless some semantic analysis of the commands is used. Automated attacks by malicious executables cannot be detected by this technique as they only analyze the keystrokes.

IDES [17] used expert system methods for misuse intrusion detection and statistical methods for anomaly detection. IDES expert system component evaluates audit records as they are produced. The audit records are viewed as facts, which map to rules in the rule-base. Firing a rule increases the suspicion rating of the user corresponding to that record. Each user's suspicion rating starts at zero and is increased with each suspicious record. Once the suspicion rating surpasses a pre-defined threshold, an intrusion is detected. There are some disadvantages to expert system method. An Intrusion scenario that does not trigger a rule will not be detected by the rule-based approach. Maintaining and updating a complex rule-based system can be difficult. The rules in the expert system have to be formulated by a security professional which means the system strength is dependent on the ability of the security personnel.

Model-Based approach attempts to model intrusions at a higher level of abstraction than audit trail records. This allows administrators to generate their representation of the penetration abstractly, which shifts the burden of determining what audit records are part of a suspect sequence to the expert system. This technique differs from the rule-based expert system technique, which simply attempt to pattern match audit records to expert rules. Garvey and Lunt's [8] model-based approach consists of three parts: anticipator, planner and interpreter. The anticipator generates the next set of behaviors to be verified in the audit trail based on the current active models and passes these sets to the planner. The planner determines how the hypothesized behavior is reflected in the audit data and translates it into a system dependent audit trail match. The interpreter then searches for this data in the audit trail. The system collects the information this way until a threshold is reached, then it signals an intrusion attempt. The advantage of this model is it can predict the intruder's next move based on the intrusion model, which is used to take preventive measures, what to look for next and verify against the intrusion hypothesis. This also reduces the data to be processed as the planner and interpreter filter the data based on their knowledge what to look for, which leads to efficiency. There are some drawbacks to this system. The intrusion patterns must always occur in the behavior it is looking for otherwise it cannot detect them.

The Pattern Matching [14] approach encodes known intrusion signatures as patterns that are then matched against the audit data. Intrusion signatures are classified using structural interrelationships among the elements of the signatures. The patterned signatures are matched against the audit trails and any matched pattern can be detected as an intrusion. Intrusions can be understood and characterized in terms of the structure of events needed to detect them. A Model of pattern matching is implemented using colored petri nets in IDIOT [15]. Intrusion signature is represented with Petri nets, the start state and final state notion is used to define matching to detect the intrusion. This system has several advantages. The system can be clearly separated into different parts. This makes different solutions to be substituted for each component without changing the overall structure of the system. Pattern specifications are declarative, which means pattern representation of intrusion signatures can be specified by defining what needs to be matched than how it is matched. Declarative specification of intrusion patterns enables them to be exchanged across different operating systems with different audit trails. There are few problems in this approach. Constructing patterns from attack scenarios is a difficult problem and it needs human expertise. Attack scenarios that are known and constructed into patterns by the system can only be detected, which is the common problem of misuse detection.

3.4 Evolving IDS Using Genetic Programming (GP)

This section provides an introduction to the two GP techniques used namely Linear Genetic Programming (LGP) and Multi Expression Programming (MEP).

3.4.1 Linear Genetic Programming (LGP)

Linear genetic programming is a variant of the GP technique that acts on linear genomes [4]. Its main characteristics in comparison to tree-based GP lies in that the evolvable units are not the expressions of a functional programming language (like LISP), but the programs of an imperative language (like C/C++). An alternate approach is to evolve a computer program at the machine code level, using lower level representations for the individuals. This can tremendously hasten the evolution process as, no matter how an individual is initially represented, finally it always has to be represented as a piece of machine code, as fitness evaluation requires physical execution of the individuals. The basic unit of evolution here is a native machine code instruction that runs on the floating-point processor unit (FPU). Since different instructions may have different sizes, here instructions are clubbed up together to form instruction blocks of 32 bits each. The instruction blocks hold one or more native machine code instructions, depending on the sizes of the instructions. A crossover point can occur only between instructions and is prohibited from occurring within an instruction. However the mutation operation does not have any such restriction.

The settings of various linear genetic programming system parameters are of utmost importance for successful performance of the system. The population space has been subdivided into multiple subpopulation or demes. Migration of individuals among the sub-populations causes evolution of the entire population. It helps to maintain diversity in the population, as migration is restricted among the demes. Moreover, the tendency towards a bad local minimum in one deme can be countered by other demes with better search directions. The various LGP search parameters are the mutation frequency, crossover frequency and the reproduction frequency: The crossover operator acts by exchanging sequences of instructions between two tournament winners. Steady state genetic programming approach was used to manage the memory more effectively.

3.4.2 Multi Expression Programming (MEP)

A GP chromosome generally encodes a single expression (computer program). By contrast, a Multi Expression Programming (MEP) chromosome encodes several expressions. The best of the encoded solution is chosen to represent the chromosome (by supplying the fitness of the individual). The MEP chromosome has some advantages over the single-expression chromosome especially when the complexity of the target expression is not known. This feature also acts as a provider of variable-length expressions. Other techniques (such as Grammatical Evolution (GE) [27] or Linear Genetic Programming (LGP) [4]) employ special genetic operators (which insert or remove chromosome parts) to achieve such a complex functionality. Multi Expression Programming (MEP) technique ([20], [21]) description and features are presented in what follows.

3.4.3 Solution Representation

MEP genes are (represented by) substrings of a variable length. The number of genes per chromosome is constant. This number defines the length of the chromosome. Each gene encodes a terminal or a function symbol. A gene that encodes a function includes pointers towards the function arguments. Function arguments always have indices of lower values than the position of the function itself in the chromosome.

The proposed representation ensures that no cycle arises while the chromosome is decoded (phenotypically transcripted). According to the proposed representation scheme, the first symbol of the chromosome must be a terminal symbol. In this way, only syntactically correct programs (MEP individuals) are obtained.

An example of chromosome using the sets $F= \{+, *\}$ and $T= \{a, b, c, d\}$ is given below:

1: a
2: b
3: + 1, 2
4: c
5: d
6: + 4, 5
7: * 3, 6

The maximum number of symbols in MEP chromosome is given by the formula:

$Number\_of\_Symbols = (n+1) * (Number\_of\_Genes - 1) + 1$,

where n is the number of arguments of the function with the greatest number of arguments.

The maximum number of effective symbols is achieved when each gene (excepting the first one) encodes a function symbol with the highest number of arguments. The minimum number of effective symbols is equal to the number of genes and it is achieved when all genes encode terminal symbols only.

The translation of a MEP chromosome into a computer program represents the phenotypic transcription of the MEP chromosomes. Phenotypic translation is obtained by parsing the chromosome top-down. A terminal symbol specifies a simple expression. A function symbol specifies a complex expression obtained by connecting the operands specified by the argument positions with the current function symbol.

For instance, genes 1, 2, 4 and 5 in the previous example encode simple expressions formed by a single terminal symbol. These expressions are:

$E_1 = a$,
$E_2 = b$,
$E_4 = c$,
$E_5 = d$,

Gene 3 indicates the operation + on the operands located at positions 1 and 2 of the chromosome. Therefore gene 3 encodes the expression: $E_3 = a + b$. Gene 6 indicates the operation + on the operands located at positions 4 and 5. Therefore gene 6 encodes the expression: $E_6 = c + d$. Gene 7 indicates the operation * on the operands located at position 3 and 6. Therefore gene 7 encodes the expression: $E_7 = (a + b) * (c + d)$. E_7 is the expression encoded by the whole chromosome.

There is neither practical nor theoretical evidence that one of these expressions is better than the others. This is why each MEP chromosome is allowed to encode a number of expressions equal to the chromosome length (number of genes). The chromosome described above encodes the following expressions:

$E_1 = a,$
$E_2 = b,$
$E_3 = a + b,$
$E_4 = c,$
$E_5 = d,$
$E_6 = c + d,$
$E_7 = (a + b) * (c + d).$

The value of these expressions may be computed by reading the chromosome top down. Partial results are computed by dynamic programming and are stored in a conventional manner.

Due to its multi expression representation, each MEP chromosome may be viewed as a forest of trees rather than as a single tree, which is the case of Genetic Programming.

3.4.4 Fitness Assignment

As MEP chromosome encodes more than one problem solution, it is interesting to see how the fitness is assigned.

The chromosome fitness is usually defined as the fitness of the best expression encoded by that chromosome.

For instance, if we want to solve symbolic regression problems, the fitness of each sub-expression E_i may be computed using the formula:

$$f(E_i) = \sum_{k=1}^{n} |o_{k,i} - w_k|,$$

where $o_{k,i}$ is the result obtained by the expression E_i for the fitness case k and w_k is the targeted result for the fitness case k. In this case the fitness needs to be minimized.

The fitness of an individual is set to be equal to the lowest fitness of the expressions encoded in the chromosome:

When we have to deal with other problems, we compute the fitness of each sub-expression encoded in the MEP chromosome. Thus, the fitness of the entire individual is supplied by the fitness of the best expression encoded in that chromosome.

3.5 Machine Learning Techniques

For illustrating the capabilities of GP systems, two popular machine learning techniques were used namely Decision Trees (DT) and Support Vector Machines (SVM).

3.5.1 Decision Trees

Decision tree induction is one of the classification algorithms in data mining. The classification algorithm is inductively learned to construct a model from the pre-classified data set. Each data item is defined by values of the attributes. Classification may be viewed as mapping from a set of attributes to a particular class. The decision tree classifies the given data item using the values of its attributes. The decision tree is initially constructed from a set of pre-classified data. The main approach is to select the attributes, which best divides the data items into their classes. According to the values of these attributes the data items are partitioned. This process is recursively applied to each partitioned subset of the data items. The process terminates when all the data items in current subset belongs to the same class. A node of a decision tree specifies an attribute by which the data is to be partitioned. Each node has a number of edges, which are labeled according to a possible value of the attribute in the parent node. An edge connects either two nodes or a node and a leaf. Leaves are labeled with a decision value for categorization of the data.

Induction of the decision tree uses the training data, which is described in terms of the attributes. The main problem here is deciding the attribute, which will best partition the data into various classes. The ID3 algorithm [25] uses the information theoretic approach to solve this problem. Information theory uses the concept of entropy, which measures the impurity of a data items. The value of entropy is small when the class distribution is uneven, that is when all the data items belong to one class. The entropy value is higher when the class distribution is more even, that is when the data items have more classes. Information gain is a measure on the utility of each attribute in classifying the data items. It is measured using the entropy value. Information gain measures the decrease of the weighted average impurity (entropy) of the attributes compared with the impurity of the complete set of data items. Therefore, the attributes with the largest information gain are considered as the most useful for classifying the data items.

To classify an unknown object, one starts at the root of the decision tree and follows the branch indicated by the outcome of each test until a leaf node is reached. The name of the class at the leaf node is the resulting classification. Decision tree induction has been implemented with several algorithms. Some of them are ID3 [25] and later on it was extended into algorithms C4.5 [26] and C5.0. Another algorithm for decision trees is CART [5].

3.5.2 Support Vector Machines (SVMs)

Support Vector Machines [31] combine several techniques from statistics, machine learning and neural networks. SVM perform structural risk minimization. They create a classifier with minimized VC (Vapnik and Chervonenkis) dimension. If the VC Dimension is low, the expected probability of error is low as well, which means good generalization. SVM has the common capability to separate the classes in the linear way. However, SVM also has another specialty that it is using a linear separating hyperplane to create a classifier, yet some problems can't be linearly separated in the original input space. Then SVM uses one of the most important ingredients called kernels, i.e., the concept of transforming linear algorithms into nonlinear ones via a map into feature spaces.

The possibility of using different kernels allows viewing learning methods like Radial Basis Function Neural Network (RBFNN) or multi-layer Artificial Neural Networks (ANN) as particular cases of SVM despite the fact that the optimized criteria are not the same [16]. While ANNs and RBFNN optimizes the mean squared error dependent on the distribution of all the data, SVM optimizes a geometrical criterion, which is the margin and is sensitive only to the extreme values and not to the distribution of the data into the feature space. The SVM approach transforms data into a feature space F that usually has a huge dimension. It is interesting to note that SVM generalization depends on the geometrical characteristics of the training data, not on the dimensions of the input space. Training a support vector machine (SVM) leads to a quadratic optimization problem with bound constraints and one linear equality constraint. Vapnik [31] shows how training a SVM for the pattern recognition problem leads to the following quadratic optimization problem:

$$\text{Minimize: } W(\alpha) = -\sum_{i-1}^{l} \alpha_i + \frac{1}{2}\sum_{i=1}^{l}\sum_{j=1}^{l} y_i y_j \alpha_i \alpha_j k(x_i, x_j)$$

$$\text{Subject to } \sum_{i=1}^{l} y_i \alpha_i, \quad \forall i : 0 \le \alpha_i \le C \quad ,$$

where l is the number of training examples $\alpha$is a vector of $l$variables and each component $\alpha_i$corresponds to a training example (x_i, y_i). SVM Torch software was used for simulating the SVM learning algorithm for IDS.

3.6 Experiment Setup and Results

The data for our experiments was prepared by the 1998 DARPA intrusion detection evaluation program by MIT Lincoln Labs [19]. The data set has 41 attributes for each connection record plus one class label as given in Table 3.1. The data set contains 24 attack types that could be classified into four main categories:

DoS: Denial of Service. Denial of Service (DoS) is a class of attack where an attacker makes a computing or memory resource too busy or too full to handle legitimate requests, thus denying legitimate users access to a machine.

Table 3.1. Variables for intrusion detection data set

Variable No.	Variable name	Variable type	Variable label
1	duration	continuous	A
2	protocol_type	discrete	B
3	service	discrete	C
4	flag	discrete	D
5	src_bytes	continuous	E
6	dst_bytes	continuous	F
7	land	discrete	G
8	wrong_fragment	continuous	H
9	urgent	continuous	I
10	hot	continuous	J
11	num_failed_logins	continuous	K
12	logged_in	discrete	L
13	num_compromised	continuous	M
14	root_shell	continuous	N
15	su_attempted	continuous	O
16	num_root	continuous	P
17	num_file_creations	continuous	Q
18	num_shells	continuous	R
19	num_access_files	continuous	S
20	num_outbound_cmds	continuous	T
21	is_host_login	discrete	U
22	is_guest_login	discrete	V
23	count	continuous	W
24	srv_count	continuous	X
25	serror_rate	continuous	Y
26	srv_serror_rate	continuous	X
27	rerror_rate	continuous	AA
28	srv_rerror_rate	continuous	AB
29	same_srv_rate	continuous	AC
30	diff_srv_rate	continuous	AD
31	srv_diff_host_rate	continuous	AE
32	dst_host_count	continuous	AF
33	dst_host_srv_count	continuous	AG
34	dst_host_same_srv_rate	continuous	AH
35	dst_host_diff_srv_rate	continuous	AI
36	dst_host_same_src_port_rate	continuous	AJ
37	dst_host_srv_diff_host_rate	continuous	AK
38	dst_host_serror_rate	continuous	AL
39	dst_host_srv_serror_rate	continuous	AM
40	dst_host_rerror_rate	continuous	AN
41	dst_host_srv_rerror_rate	continuous	AO

R2L: Unauthorized Access from a Remote Machine. A remote to user (R2L) attack is a class of attack where an attacker sends packets to a machine over a network, then exploits the machine's vulnerability to illegally gain local access as a user.

U2Su: Unauthorized Access to Local Super User (root). User to root (U2Su) exploits are a class of attacks where an attacker starts out with access to a normal user account on the system and is able to exploit vulnerability to gain root access to the system.

Probing: Surveillance and Other Probing. Probing is a class of attack where an attacker scans a network to gather information or find known vulnerabilities. An attacker with a map of machines and services that are available on a network can use the information to look for exploits.

Experiments presented in this chapter have two phases namely training and testing. In the training phase, MEP/LGP models were constructed using the training data to give maximum generalization accuracy on the unseen data. The test data is then passed through the saved trained model to detect intrusions in the testing phase. The 41 features are labeled as shown in Table 3.1 and the class label is named as *AP*.

This data set has five different classes namely *Normal, DoS, R2L, U2R* and *Probes.* The training and test comprises of 5,092 and 6,890 records respectively [13]. All the training data were scaled to (0-1). Using the data set, we performed a 5-class classification. The normal data belongs to class 1, probe belongs to class 2, denial of service belongs to class 3, user to super user belongs to class 4, remote to local belongs to class 5.

The settings of various linear genetic programming system parameters are of utmost importance for successful performance of the system [1]. The various parameter setting for LGP is depicted in Table 3.2

Table 3.2. Parameter settings for LGP

Parameter	Normal	Probe	DoS	U2Su	R2L
Population size	2048	2048	2048	2048	2048
Maximum no of tournaments	120000	120000	120000	120000	120000
Tournament size	8	8	8	8	8
Mutation frequency (%)	85	82	75	86	85
Crossover frequency (%)	75	70	65	75	70
Number of demes	10	10	10	10	10
Maximum program size	256	256	256	256	256

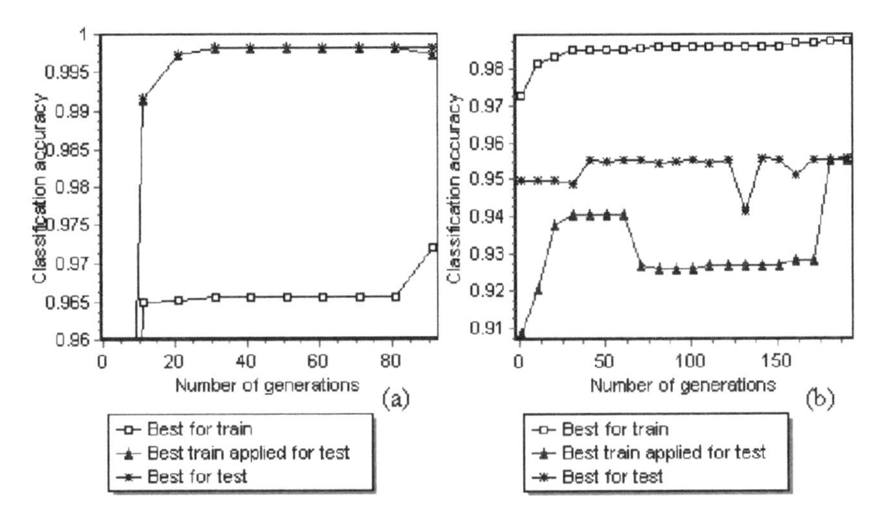

Fig. 3.3. Relation between accuracy and number of generations: (a) normal mode (b) probe attacks

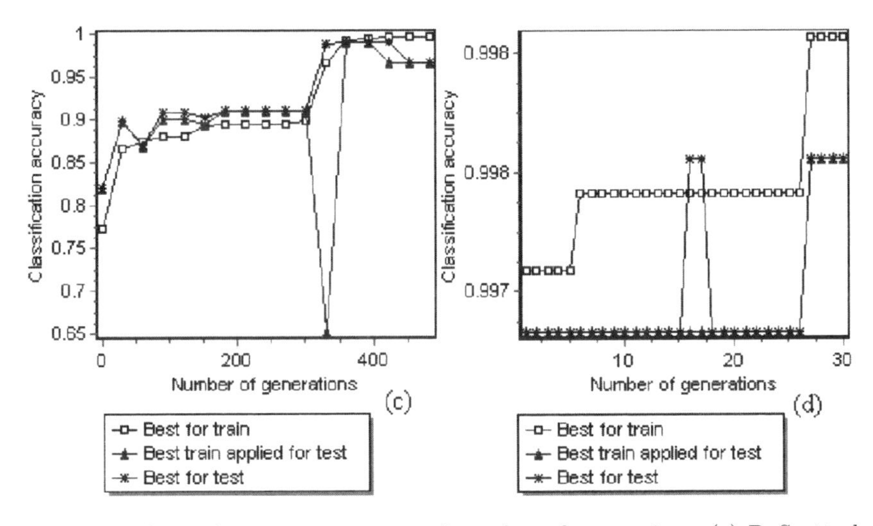

Fig. 3.4. Relation between accuracy and number of generations: (c) DoS attacks (d) U2R attacks

Our trial experiments with SVM revealed that the polynomial kernel option often performs well on most of the attack classes. We also constructed decision trees using the training data and then testing data was passed through the constructed classifier to classify the attacks [22].

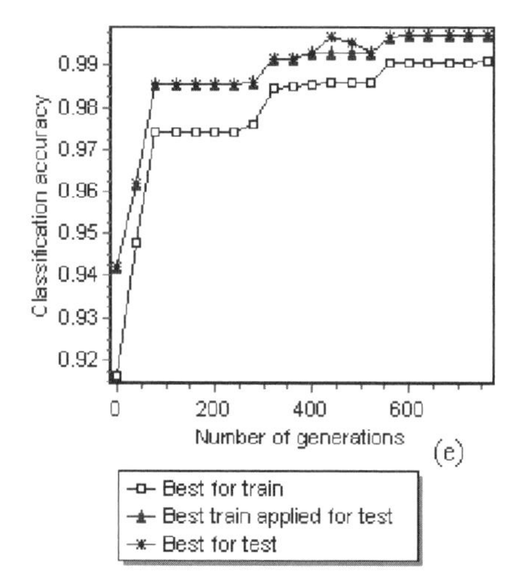

Fig. 3.5. Relation between accuracy and number of generations: (e) R2L attacks

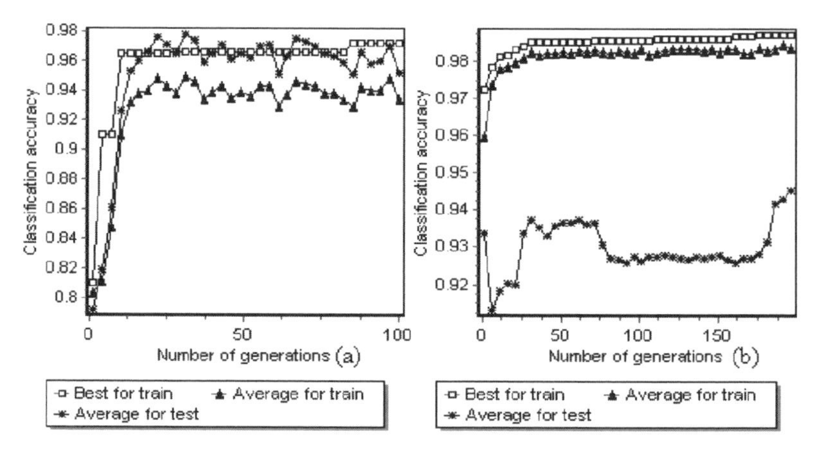

Fig. 3.6. Relationship between the best result obtained for training data set and the average of results obtained for training /test data: (a) normal mode (b) probe attacks

Parameters used by MEP are presented in Table 3.3 [9]. We made use of $+$, $-$, $*$, $/$, sin, cos, *sqrt*, *ln*, *lg*, $\log_2$, min, max, and *abs* as function sets.

Experiment results (for test data set) using the four techniques are depicted in Table 3.4. In Table 3.5 the variable combinations evolved by MEP are presented.

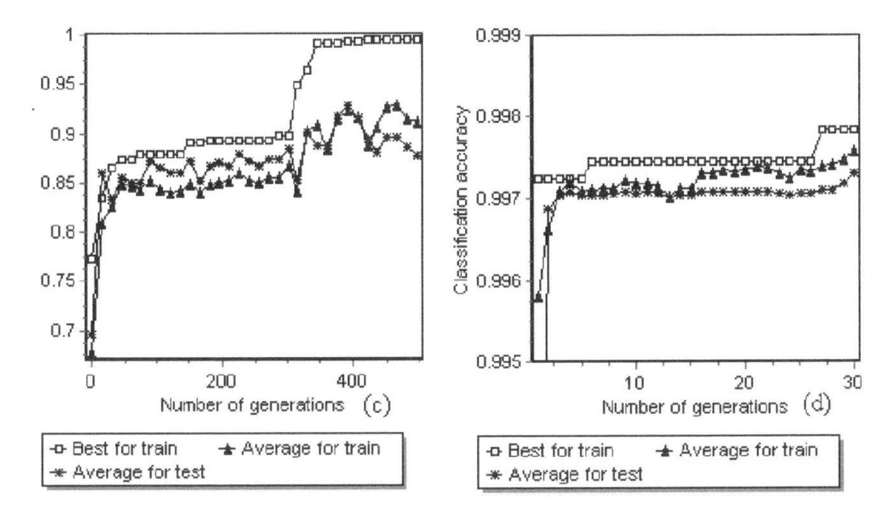

Fig. 3.7. Relationship between the best result obtained for training data set and the average of results obtained for training /test data: (c) DoS attacks (d) U2R attacks

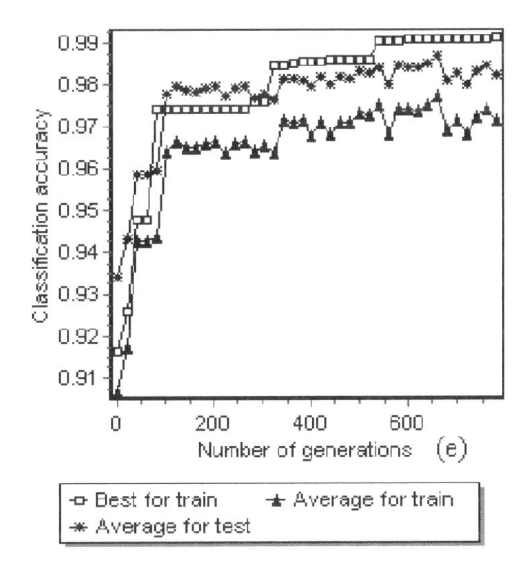

Fig. 3.8. Relationship between the best result obtained for training data set and the average of results obtained for training /test data: (e) R2L attacks

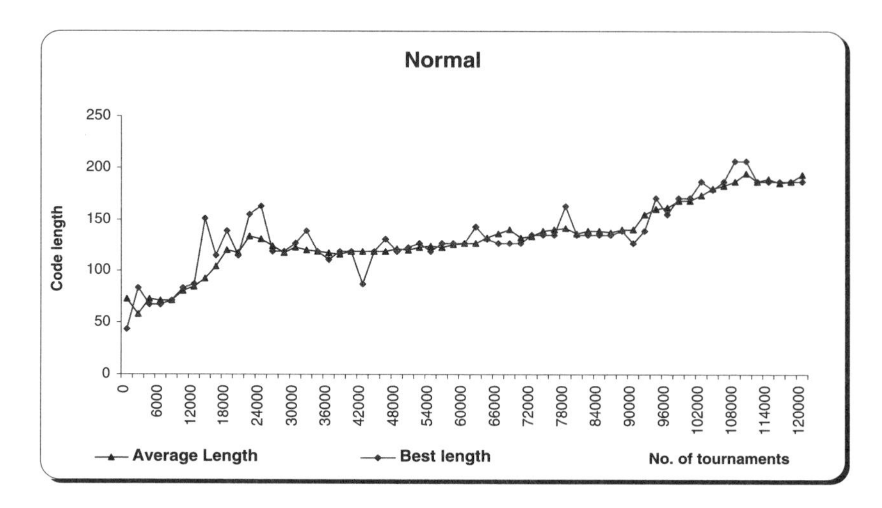

Fig. 3.9. Growth of program codes for normal mode

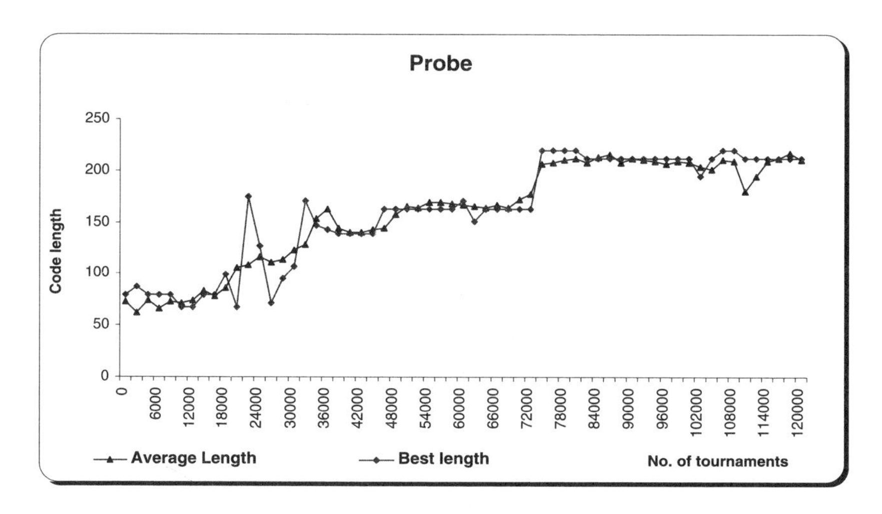

Fig. 3.10. Growth of program codes for probe attacks

As evident from Table 3.5, the presented GP techniques out performed some of the most popular machine learning techniques. MEP performed well for Classes 1, 4 and 5 while LGP gave the best test results Classes 2 and 3.

MEP performance is illustrated in Figures 3.3, 3.4, 3.5, 3.6, 3.7, and 3.8. The classification accuracy for the best results obtained for training data, average of results obtained for the test data using the best training function and the best results obtained for the test data are depicted. Figures 3.3 (a) and (b) correspond to Normal, and Probe attacks, Figures 3.4 (c) and (d)

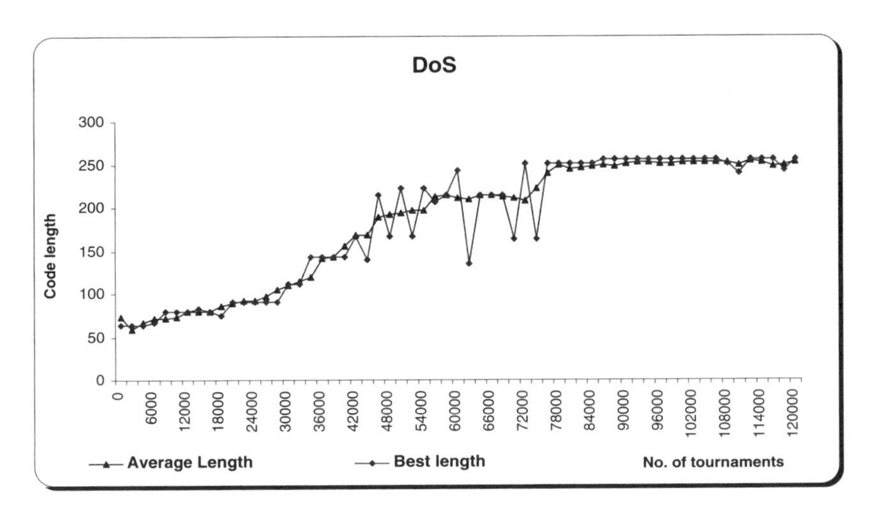

Fig. 3.11. Growth of program codes for DOS attacks

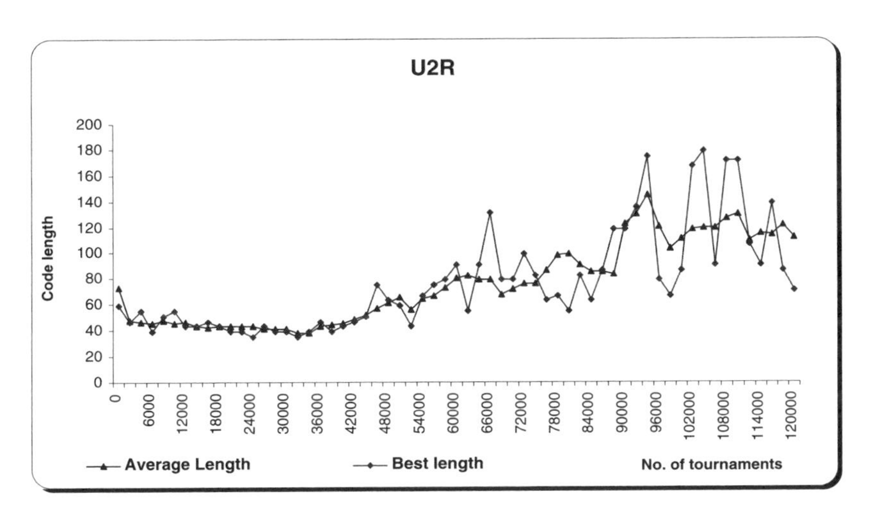

Fig. 3.12. Growth of program codes for U2R attacks

corresponds to DOS and U2R attacks and Figure 3.5 (e) corresponds to R2L attacks respectively.

In Figures 3.6- 3.8, the average of the classification accuracies for the results best results obtained for training data, results obtained for the test data and the obtained for the training data are depicted. Figures 3.6 (a) and (b) corresponds to Normal and Probe attacks, Figures 3.7 (c) and (d) corresponds DOS and U2R attacks and Figure 3.8 (e) corresponds to R2L attacks respectively.

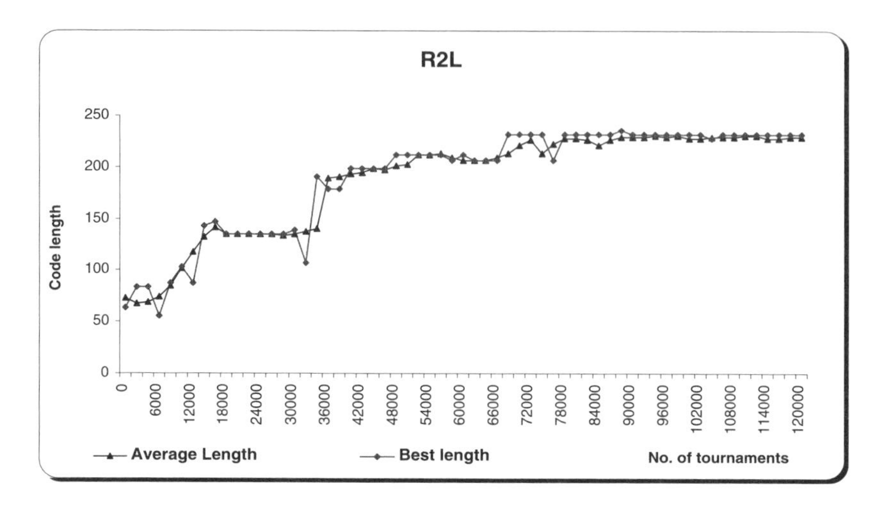

Fig. 3.13. Growth of program codes for R2L attacks

Table 3.3. Parameters used by MEP

Attack type	Parameter value				
	Pop. Size	Generations	Crossover (%)	No. of mutations	Chromosome length
Normal	100	30	0.9	3	30
Probe	200	200	0.9	4	40
DOS	250	800	0.8	5	40
U2R	100	20	0.9	3	30
R2L	100	800	0.9	4	40

Table 3.4. Functions evolved by MEP

Attack type	Evolved Function
Normal	$var12 * log_2(var10 + var3)$
Probe	$(fabs(var30 + var35)) < (var26 + var27)?(fabs(var30 + var35)) : (var26 + var27);$
DOS	$return(var38 - (Ln(var41 * var6) + sin(Lg(var30))) - (Lg(var30) - (var41 * var6))) > (0.3415 + var24 + var41 * var6)?(var38 - (Ln(var41 * var6) + sin(Lg(var30))) - (Lg(var30) - (var41 * var6))) : (0.3415 + var24 + var41 * var6) + var8$
U2R	$sin(var14) - var33$
R2L	$fabs((fabs(var8 > (var1 + (var6 > (Ln(var6))?var6 : (Ln(var6))) * var3)?var10 : (var1 + (var6 > (Ln(var6))?var6 : (Ln(var6))) * var3))) * (var12 + var6)) + var11$

Table 3.5. Performance comparison

Attack type	Classification accuracy on test data set (%)			
	MEP	DT	SVM	LGP
Normal	**99.82**	99.64	99.64	99.73
Probe	95.39	99.86	98.57	**99.89**
DOS	98.94	96.83	99.92	**99.95**
U2R	**99.75**	68.00	40.00	64.00
R2L	**99.75**	84.19	33.92	99.47

Figures 3.9 - 3.13 illustrate the growth in the program codes during the 120 tournaments during the development of LGP models. The best and average code length is depicted during the evolutionary learning.

In some classes the accuracy figures tend to be very small and may not be statistically significant, especially in view of the fact that the 5 classes of patterns differ in their sizes tremendously. For example only 27 data sets were available for training the U2R class. More definitive conclusions can only be made after analyzing more comprehensive sets of network traffic.

3.7 Conclusions

This chapter illustrated the importance of GP techniques for evolving intrusion detection systems. MEP outperforms LGP for three of the considered classes and LGP outperform MEP for two of the classes. MEP classification accuracy is grater than 95% for all considered classes and for three of them is greater than 99.75%. It is to be noted that for real time intrusion detection systems MEP and LGP would be the ideal candidates because of its simplified implementation.

Acknowledgements

This research was supported by the International Joint Research Grant of the IITA (Institute of Information Technology Assessment) foreign professor invitation program of the MIC (Ministry of Information and Communication), South Korea.

References

1. Abraham, A., Evolutionary Computation in Intelligent Web Management, Evolutionary Computing in Data Mining, Ghosh A. and Jain L.C. (Eds.), Studies in Fuzziness and Soft Computing, Springer Verlag Germany, Chapter 8, pp. 189-210, 2004.

2. J. P. Anderson. Computer Security Threat Monitoring and Surveillance. Technical report, James P Anderson Co., Fort Washington, Pennsylvania, April 1980.
3. Barbara D., Couto J., Jajodia S. and Wu N., ADAM: A Testbed for Exploring the Use of Data Mining in Intrusion Detection. SIGMOD Record, 30(4), pp. 15-24, 2001.
4. Brameier M. and Banzhaf W, Explicit control of diversity and effective variation distance in Linear Genetic Programming. In Proceedings of the fourth European Conference on Genetic Programming, Springer-Verlag Berlin, 2001.
5. Brieman L., Friedman J., Olshen R., and Stone C., Classification of Regression Trees. Wadsworth Inc., 1984.
6. Cohen W., Learning Trees and Rules with Set-Valued Features, American Association for Artificial Intelligence (AAAI), 1996.
7. Denning D., An Intrusion-Detection Model, IEEE Transactions on Software Engineering, Vol. SE-13, No. 2, pp. 222-232, 1987.
8. T. D. Garvey and T. F. Lunt. Model based intrusion detection, In Proceedings of the 14th National Computer Security Conference, pages 372-385, October 1991.
9. Grosan C., Abraham A. and Han S.Y., MEPIDS: Multi-Expression Programming for Intrusion Detection System, International Work-conference on the Interplay between Natural and Artificial Computation, (IWINAC'05), Spain, Lecture Notes in Computer Science, Springer Verlag, Germany, pp. 163-172, 2005.
10. R. Heady, G. Luger, A. Maccabe, and M. Servilla, The Architecture of a Network level Intrusion Detection System. Technical report, Department of Computer Science, University of New Mexico, August 1990.
11. K. Ilgun. USTAT: A Real-Time Intrusion Detection System for UNIX, Master Thesis, University of California, Santa Barbara, November 1992.
12. T. Joachims. Making Large-Scale SVM Learning Practical. LS8-Report, University of Dortmund, LS VIII-Report, 1998.
13. KDD Cup 1999 Intrusion detection data set: http://kdd.ics.uci.edu/databases/kddcup99/kddcup.data_10_percent.gz
14. S. Kumar and E. H. Spafford. An Application of Pattern Matching in Intrusion Detection. Technical Report CSD-TR-94-013, Purdue University, 1994.
15. S. Kumar. Classification and Detection of Computer Intrusions, PhD Thesis, Department of Computer Science, Purdue University, August 1995.
16. Lee W. and Stolfo S. and Mok K., A Data Mining Framework for Building Intrusion Detection Models. In Proceedings of the IEEE Symposium on Security and Privacy, 1999.
17. T.F. Lunt, A. Tamaru, F. Gilham et al, A Real Time Intrusion Detection Expert System (IDES), Final Technical Report, Project 6784, SRI International 1990
18. T. Lunt. Detecting intruders in computer systems. In Proceedings of the 1993 Conference on Auditing and Computer Technology, 1993.
19. MIT Lincoln Laboratory. http://www.ll.mit.edu/IST/ideval/
20. Oltean M. and Grosan C., A Comparison of Several Linear GP Techniques, *Complex Systems*, Vol. 14, No. 4, pp. 285-313, 2004.
21. Oltean M. and Grosan C., Evolving Evolutionary Algorithms using Multi Expression Programming. *Proceedings of The 7th European Conference on Artificial Life*, Dortmund, Germany, pp. 651-658, 2003.
22. Peddabachigari S., Abraham A., Thomas J., Intrusion Detection Systems Using Decision Trees and Support Vector Machines, International Journal of Applied Science and Computations, Vol.11, No.3, pp.118-134, 2004. .

23. P. A. Porras. STAT: A State Transition Analysis Tool for Intrusion Detection. Master's Thesis, Computer Science Dept., University of California, Santa Barbara, 1992.
24. Provost, F. and T. Fawcett. Robust Classification for Imprecise Environments, Machine Learning 42, 203-231, 2001.
25. J. R. Quinlan. Induction of Decision Trees. Machine Learning, 1:81-106, 1986.
26. J. R. Quinlan. C4.5: Programs for Machine Learning. Morgan Kaufmann, 1993.
27. C. Ryan C. J.J. Collins and M. O'Neill. Gramatical Evolution: Evolving programs for an arbitrary language, In Proceedings of the first European Workshop on Genetic Programming, Springer-Verlag, Berlin, 1998.
28. Summers R.C., Secure Computing: Threats and Safeguards. McGraw Hill, New York, 1997.
29. A. Sundaram. An Introduction to Intrusion Detection. ACM Cross Roads, Vol. 2, No. 4, April 1996.
30. H. S. Teng, K. Chen and S. C. Lu. Security Audit Trail Analysis Using Inductively Generated Predictive Rules. In Proceedings of the 11th National Conference on Artificial Intelligence Applications, pages 24-29, IEEE, IEEE Service Center, Piscataway, NJ, March 1990.
31. Vapnik V.N., The Nature of Statistical Learning Theory. Springer, 1995.

4

Evolutionary Pattern Matching Using Genetic Programming

Nadia Nedjah[1] and Luiza de Macedo Mourelle[2]

[1] Department of Electronics Engineering and Telecommunications,
Engineering Faculty,
State University of Rio de Janeiro,
Rua São Francisco Xavier, 524, Sala 5022-D,
Maracanã, Rio de Janeiro, Brazil
nadia@eng.uerj.br, http://www.eng.uerj.br/~nadia

[2] Department of System Engineering and Computation,
Engineering Faculty,
State University of Rio de Janeiro,
Rua São Francisco Xavier, 524, Sala 5022-D,
Maracanã, Rio de Janeiro, Brazil
ldmm@eng.uerj.br, http://www.eng.uerj.br/~ldmm

Pattern matching is a fundamental feature in many applications such as functional programming, logic programming, theorem proving, term rewriting and rule-based expert systems. Usually, patterns size is not constrained and ambiguous patterns are allowed. This generality leads to a clear and concise programming style. However, it yields challenging problems in compiling of such programming languages. In this chapter, patterns are pre-processed into a deterministic finite automaton.

With ambiguous or overlapping patterns a subject term may be an instance of more than one pattern. In this case, pattern matching order in lazy evaluation affects the size of the matching automaton and the matching time. Furthermore, it may even impact on the termination properties of term evaluations.

In this chapter, we engineer good traversal orders that allow one to design an efficient adaptive pattern-matchers that visit necessary positions only. We do so using genetic programming to evolve the most adequate traversal order given the set of allowed patterns. Hence, we improve time and space requirements of pattern-matching as well as termination properties of term evaluation.

N. Nedjah and L. de M. Mourelle: *Evolutionary Pattern Matching Using Genetic Programming*, Studies in Computational Intelligence (SCI) **13**, 81–104 (2006)
www.springerlink.com

4.1 Introduction

Pattern matching is an important operation in several applications such as functional, equational and logic programming [5, 18], theorem proving [4] and rule-based expert systems [3]. With ambiguous patterns, an input term may be an instance of more than one pattern. Usually, patterns are partially ordered using priorities.

Pattern-matching automata have been studied for over a decade. Pattern-matching can be achieved as in lexical analysis by using a finite automaton [2, 7, 8, 13, 20]. Gräf [7] and Christian [2] construct deterministic matching automata for unambiguous patterns based on the left-to-right traversal order. In functional programming, Augustsson [1] and Wadler [22] describe matching techniques that are also based on left-to right traversal of terms but allow prioritised overlapping patterns. Although these methods are economical in terms of space usage, they may re-examine symbols in the input term. In the worst case, they can degenerate to the naive method of checking the subject term against each pattern individually. In contrast, Christian's [2] and Gräf's [7] methods avoid symbol re-examination at the cost of increased space requirements. In order to avoid backtracking over symbols already examined, like Gräf's our method introduces new patterns. These correspond to overlaps in the scanned prefixes of original patterns. When patterns overlap, some of the added patterns may be irrelevant to the matching process. In previous work [15], we proposed a method that improves Gräf's in the sense that it introduces only a subset of the patterns that his method adds. This improves both space and time requirements as we will show later. Sekar [20] uses the notion of irrelevant patterns to compute traversal orders of pattern-matching. His algorithm eliminates the pattern π whenever its match implies a match for a pattern of higher priority than π. In contrast with Sekar's method, we do not introduce irrelevant patterns at once [15].

The pattern-matching order in lazy evaluation may affect the size of the matching automaton, the matching time and in the worst case, the termination properties of term evaluations [22]. The adaptive strategy is the top-down left-to-right lazy strategy used in most lazy functional languages [20]. It selects the leftmost-outermost redex but may force the reduction of a subterm if the root symbol of that subterm fails to match a function symbol in the patterns. So, the order of such reductions coincides with that of pattern-matching. Using left-to-right pattern matching, a subject term evaluation may fail to terminate only because of forcing reductions of subterms when it is unnecessary before declaring a match. For the left-to-right traversal order, such unnecessary reductions are required to ensure that no backtracking is needed when pattern-matching fails.

In this chapter, we study strategies to select a traversal order that should normally improve space requirements and/or matching times. Indexes for a pattern are positions whose inspection is necessary to declare a match. Inspecting them first for patterns which are not essentially strongly sequential allows us to engineer adaptive traversal orders that should improve space usage and matching times as shown in [11, 20]. When no index can be found for a pattern set, choosing the best next position to reduce and then match is *NP*-problem. We conclude the chapter by engineering good traversal orders for some benchmarks to prove the efficiency of the genetic programming-based approach.

The rest of the chapter is organised as follows: In Section 4.2, we define some notation and necessary terminologies. Following that, in Section 4.3, we state the problem of pattern matching and the impact that the traversal order of the patterns has on the process efficiency, when the patterns are ambiguous. Consequently, in Section 4.4, we introduce some heuristics that allows one engineer a relatively good traversal order. Thereafter, in Section 4.5, we show how the traversal order engineering problem may be approached using genetic programming. We define the encoding of traversal orders and the corresponding decoding into adaptive pattern-matchers. Later, in the same section, we develop the necessary genetic operators and we engineer the fitness function that will evaluate how good is the evolved traversal order. Finally, in Section 4.6, we present some interesting results obtained genetically and compare them to their counterparts from [15].

4.2 Preliminary Notation and Terminology

Definition 1. An equational program can be defined as a 4-tuple $EP = \langle F, V, R, T \rangle$, where $F = \{f, g, h, \ldots\}$ is a set of function symbols, $V = \{x, y, z, \ldots\}$ is a set of variable symbols and $R = \{\pi_1 \rightarrow tau_1, \pi_2 \rightarrow tau_2, \ldots\}$ is a set of rewrite rules called the term rewriting system, where π_i and $\rightarrow_i$ are terms, called the pattern and template respectively. T is the subject term, which is the expression to evaluate.

For convenience, in most of the paper, we will consider the patterns to be written from the symbols in $F \cup \{\omega\}$, where ω is a meta-symbol used whenever the symbol representing a variable does not matter. For a pattern set Π, we denote by F_Π the subset of F containing only the function symbols in the patterns of Π. A term is either a variable, a constant symbol or has the form $f(\sigma_1, \sigma_2, \ldots, \sigma_n)$, where each σ_i with $(1 \leq i \leq n)$ is itself a term and n is the arity of the function symbol f denoted by $\#f$. The subject term is supposed to be a ground term, i.e. a term containing no variable occurrences. Terms are interpreted syntactically as trees labelled with symbols from $F \cup V$. An instance of a term t can be obtained by replacing leaves labelled with variable symbols by other terms. In practice, however, both the subject term and templates are turned into directed acyclic graphs (DAGs) so that common subterms may be shared (represented physically once). This allows for the evaluation of such subterms to be performed at most once during the whole rewriting process.

Definition 2. A position in a term is a path specification which identifies a node in the graph of that term and therefore both the subterm rooted at that point and the symbol which labels that node. A position is specified here using a list of positive integers. The empty list Λ denotes the graph root, the position k denotes the kth child of the root, and the position $p.k$ denotes the kth ($k \geq 1$) child from the position given by p. The symbol, respectively subterm, rooted at position p in a term t is denoted by $t[p]$, respectively t/p. A position in a term is valid if, and only if, the term has a symbol at that position. So Λ is valid for any term and a position $p = q.k$ is valid if, and only if, the position q is valid, the symbol f at q is a function symbol and $k \leq \#f$.

For instance, in the term $t = f(g(a, h(a, a)), b(a, x), c)$, $t[\Lambda]$ denotes the single occurrence of f, $t[2.2]$ denotes the variable symbol x whereas $t[2]$ denotes the symbol b while $t/2$ indicates the subterm $b(a, x)$ and the positions 2.2.1 and 1.3 are not valid. In the following, we will abbreviate terms by removing parentheses and commas. For instance, t abbreviates to $fgahaabaac$. This will be unambiguous since the given function arities (i.e. $\#f = 3$, $\#g = \#h = \#b = 2$, $\#a = \#c = 0$) will be kept unchanged throughout all examples. In particular, the arities $\#f = 3$, $\#g = 2$ and $\#a = 0$ will be used in the running example.

Definition 3. A pattern set Π is *overlapping* if there is a ground term that is an instance of at least two distinct patterns in Π.

For instance, the set $\Pi = \{fa\omega\omega, f\omega a\omega\}$ is an overlapping pattern set because the term $faac$ is an instance of both patterns whereas the set $\Pi' = \{fa\omega\omega, fc\omega\omega\}$ is a non-overlapping pattern set. A similar notion is that of pattern prefix-overlapping:

Definition 4. A pattern set Π is *prefix-overlapping* if there is a ground term with a non-empty prefix that is an instance of prefixes of at least two distinct patterns in Π.

For instance, the set $\Pi = \{f\omega aa, f\omega\omega c\}$ is a non-overlapping pattern set but it is a prefix-overlapping because the prefix faa of the term $faaa$ is an instance of both prefixes $f\omega a$ and $f\omega\omega$.

When overlapping patterns are allowed in equational programming a meta-rule is needed to decide which rule should be matched when a conflict due to overlapping patterns arises. The meta-rule defines a priority relationship among overlapping patterns. Thus given a pattern set and a meta-rule, we can formalise the notion of pattern-matching as follows:

Definition 5. A term t matches a pattern $\pi_i \in \Pi$ if and only if t is an instance of π_i and t is not an instance of any pattern $\pi_j \in \Pi$ such that the priority of π_j is higher than that of π_i.

Definition 6. A term t_1 is more general than a term t_2 at a given common valid position p if and only if $t_1[p] \in V$, $t_2[p] \in F$ and the prefixes of t_1 and t_2 ending immediately before p are the same.

Definition 7. The closed pattern set $\overline{\Pi}$ corresponding to a given pattern set Π is the set obtained by applying to Π the closure operation defined by Grä f [7] as follows: For any $s \in F \cup \{\omega\}$, let Π/s be the set of elements of Π starting with s but with the first symbol s removed. Define Π_ω and Π_f, wherein $f \in F$ by (4.1):

$$\Pi_\omega = \Pi/\omega$$

$$\Pi_f = \begin{cases} \Pi/f \cup \omega^{\#f}\,\Pi/\omega & \text{if } \Pi/f = \emptyset \\ \\ \emptyset & \text{otherwise} \end{cases} \tag{4.1}$$

The closure operation is then defined recursively by (4.2:

$$\overline{\Pi} = \begin{cases} \Pi & \text{if } \Pi = \{\epsilon\} or \Pi = \emptyset \\ \\ \bigcup_{s \in F \cup \{\omega\}} s\overline{\Pi_s} & \text{otherwise} \end{cases} \tag{4.2}$$

Here ϵ is the empty string and $\omega^{\#f}$ is a repetition of $\#f$ symbols ω. The set Π_f includes all the patterns in Π starting with f, but with f removed. In addition, while factorising a pattern set according to a function symbol f (i.e. computing Π_f) the operation above takes account of the components starting with a variable symbol as well; the symbol ω is considered as possibly representing a subterm whose root symbol is f. Therefore, a new component is added to the set Π_f. This component is obtained by replacing ω by a sequence of ωs whose length is $\#f$. This sequence stands for the arguments of f.

The closure operation supplies the pattern set with some instances of the original patterns. In effect, if one pattern is more general than another at some position p then the pattern with ω replaced by $f\omega^{\#f}$ is added. For instance, the prefix-overlapping pattern set $\Pi = \{f\omega a\omega, f\omega\omega a, f\omega g\omega\omega g\omega\omega\}$ can be converted to an equivalent closed pattern set $\overline{\Pi}$ using the closure operation computed as follows (4.3):

$$\begin{aligned} \overline{\Pi} &= f\overline{\Pi_f} = f\overline{\{\omega a\omega, \omega\omega a, \omega g\omega\omega g\omega\omega\}} \\ &= f\omega\overline{\{a\omega, \omega a, g\omega\omega g\omega\omega\}} \\ &= f\omega\left(a\overline{\{\omega, a\}} \cup \omega\overline{\{a\}} \cup g\overline{\{\omega\omega g\omega\omega\}}\right) \\ &= f\omega\left(\{a\omega, aa, \omega a, g\omega\omega g\omega\omega, g\omega\omega a\}\right) \end{aligned} \tag{4.3}$$

Then, the closed pattern set corresponding to Π is (4.4):

$$\overline{\Pi} = \{f\omega a\omega, f\omega aa, f\omega\omega a, f\omega g\omega\omega g\omega\omega, f\omega g\omega\omega a\} \quad (4.4)$$

It is clear that the new pattern set accepts the same language Π does since the added patterns are all instances of the original ones. The closure operation terminates and can be computed incrementally (for full detailed description and formal proofs see [7]).

With closed pattern sets, if a pattern π_1 is more general than a pattern π_2 at position p, then $\pi_2[p]$ is checked first. This does not exclude a match for π_1 because the closed pattern set does contain a pattern that is π_1 with $\pi_1[p]$ replaced by $\pi_2[p]\omega^{\#\pi_2[p]}$. Under this assumption, an important property of such closed pattern sets is that they make it possible to determine whether a target term is a redex merely by scanning that term from left-to-right without backtracking over its symbols.

Symbol re-examination cannot be avoided in the case of non-closed prefix-overlapping pattern sets whatever the order in which the patterns are provided. For instance, let Π be the prefix-overlapping set $\{fc\omega c, f\omega ga\omega a\}$. Using the textual order meta-rule, the first pattern must be matched first, if possible. Then the term $fcgaaa$ cannot be identified as an instance of the second pattern without backtracking to the first occurrence of c when the last symbol a is encountered. However, the closure $\overline{\Pi} = \{fcga\omega c, fcga\omega a, fcg\omega\omega c, fc\omega c, f\omega ga\omega a\}$ allows matching without backtracking. Then the term will match the second pattern.

4.3 Adaptive Pattern Matching

States of adaptive automata are computed using *matching items* and *matching sets*. Since here the traversal order is not fixed a priori (i.e., it will be computed during the automaton construction procedure), the symbols in the patterns may be accepted in any order. When the adaptive order coincides with the left-to-right order, matching items and matching sets coincide with left-to-right matching items and sets respectively [14].

Definition 8. A *matching item* is a pattern in which all the symbols already matched are ticked, i.e. they have the check mark $\checkmark$. Moreover, it contains the matching dot $\bullet$ which only designates the matching symbol, i.e. the symbol to be checked next. The position of the matching symbol is called the *matching position*. A *final* matching item, namely one of the form $\pi\bullet$, has the *final* matching position which we write ∞. Final matching item may contain unchecked positions. These positions are irrelevant for announcing a match and so must be labelled with the symbol ω. Matching items are associated with a rule name.

The term obtained from a given item by replacing all the terms with an unticked root symbol by the placeholder $\_$ is called the context of the items. For instance, the context of the item $f^{\checkmark}\omega a \bullet \omega g\omega a a^{\checkmark}$ is the term $f(\_,\_,a)$

where the arities of f, g and a are as usual 3, 2 and 0. In fact, no symbol will be checked until all its parents are all checked. So, the positions of the placeholders in the context of an item are the positions of the subterms that have not been checked yet. The set of such positions for an item i is denoted by $up(i)$ (short for unchecked positions).

Definition 9. A *matching set* is a set of matching items that have the same context and a common matching position. The *initial* matching set contains items of the form $\bullet\pi$ because we recognise the root symbol (which occurs first) first whereas, *final* matching sets contain items of the form $\pi\bullet$, where π is a pattern. For initial matching sets, no symbol is ticked. A final matching set must contain a final matching item, i.e. in which all the unticked symbols are ωs. Furthermore, the rule associated with that item must be of highest priority amongst the items in the matching set.

Since the items in a matching set M have a common context, they all share a common list of unchecked positions and we can safely write $up(M)$. The only unchecked position for an initial matching set is clearly the empty position Λ.

4.3.1 Constructing Adaptive Automata

We describe adaptive automata by a 4-tuple $\langle S_0, S, Q, \delta \rangle$, wherein S is the state set, $S_0 \in S$ is the initial state, $Q \in S$ is the set of final states and δ is the state transition function. The transition function δ of an adaptive automaton is defined using three functions namely, *accept*, *choose*, and *close*. The function *accept* and *close* are similar to those of the same name in [14], and *choose* picks the next matching position. For each of these functions, we give an informal description followed by a formal definition, except for choose for which we present an informal description. Its formal definition will be discussed in detail in the next section.

1. *accept*: For a matching set and an unchecked position, this function accepts and ticks the symbol immediately after the matching dot in the items of the matching set and inserts the matching dot immediately before the symbol at the given unchecked position. Let t be a term in which some symbols are ticked. We denote by $t_{\bullet p}$ the matching item which is t with the matching dot inserted immediately before the symbol $t[p]$. then the definition of accept is as in (4.5):

$$accept(M, s, p) = \left\{ (\alpha s^{\checkmark} \beta)_{\bullet p} \mid \alpha \bullet s\beta \in M \right\} \tag{4.5}$$

2. *choose*: This function selects a position among those that are unchecked for the matching set obtained after accepting the symbol given. For a matching set M and a symbol s, the choose function selects the position

which should be inspected next. The function may also return the final position ∞. In general, the set of such positions is denoted by $\overline{up}(M,s)$. It represents the unchecked positions of $\delta(M,s)$, and consists of $up(M)$ with the position of the symbol s removed. Moreover, if the arity of s is positive then there are $\#s$ additional unchecked terms in the items of the set $\delta(M,s)$, assuming that ωs are of arity 0. Therefore, the positions of these terms are added. Recall that if a function symbol f is at position p in a term, then the arguments of f occur at positions $p.1$, $p.2$, $\ldots$, $p.\#f$. We can now formulate a definition for the set of unchecked positions $up(\delta(M,s)) = \overline{up}(M,s)$ as in (4.6), wherein p_s is the position of symbol s which is also the matching position of M.

$$\overline{up}(M,s) = \begin{cases} up(M) \setminus \{p_s\} & \text{if } \#s = 0 \\ \\ (up(M) \setminus \{p_s\}) \cup \{p.1, p.2, \ldots, p.\#f\} & \text{otherwise} \end{cases} \tag{4.6}$$

For the construction procedure of the left-to-right matching automaton, the *choose* function is not required because the leftmost unchecked position, if it exists, is always selected. Otherwise, the final position is used.

3. *close*: Given a matching set, this function computes its closure in the same way as for the left-to-right matching automaton. As it is shown in (4.7), the function adds an item $\alpha \bullet f\omega^{\#f}\beta$ to the given matching set M whenever an item $\alpha \bullet \omega\beta$ is in M together with at least one item of the form $\alpha \bullet f\beta'$.

$$\begin{array}{rl} close(M) = M \cup & \left\{ r : \alpha \bullet f\omega^{\#f}\beta \mid r : \alpha \bullet \omega\beta \in M \text{ and} \right. \\ & \exists \beta' \text{ s.t.} r' : \alpha \bullet f\beta' \in M \text{ and } \forall \beta' \\ & \text{If } r'' : \alpha \bullet f\beta'' \in M \text{ and } \alpha\omega\beta \prec \alpha f \beta'' \\ & \left. \text{Then } \neg \left(\alpha f \omega^{\#f}\beta \lhd \alpha f \beta'' \right) \right\} \end{array} \tag{4.7}$$

The definition of function *close* above allows us to avoid introducing irrelevant items [17] wherein $\prec$ and $\lhd$ express the priority rule and the instantiation relation between patterns respectively.
For a matching set M and a symbol $s \in F \cup \{\omega\}$, the transition function for an adaptive automaton can now be formally defined by the composition of the three functions *accept*, *choose* and *close* is $\delta(M,s) = close\,(accept\,(M, s, choose\,(M,s)))$. This function proceeds by first computing the list of unchecked positions of the expected set $\delta(M,s)$, then selecting an unchecked position p in $\overline{up}(M,s)$ and computing the matching set $N = accept(M,s,p)$ and finally computing the closure of N, i.e. $\delta(M,s) = close(N)$. The matching set $accept(M,s,p)$ is called the kernel of $\delta(M,s)$.

4.3.2 Example of Adaptive Automaton Construction

Consider the set of patterns $\Pi = \{r_1 : f\omega a\omega, r_2 : f\omega\omega a, r_3 : f\omega g\omega\omega g\omega\omega\}$. An adaptive automaton for Π using matching sets is shown in Figure 4.1. The choice of the positions to inspect will be explained next section. Transitions corresponding to failures are omitted, and the ω-transitions is only taken when no other available transition which accepts the current symbol. Notice that for each of the matching sets in the automaton, the items have a common set of symbols checked. Accepting the function symbol f from the initial matching set (i.e. that which labels state 0) yields the state $\{f^{\checkmark}\omega\bullet a\omega, f^{\checkmark}\omega\bullet\omega a, f^{\checkmark}\omega\bullet g\omega\omega g\omega\omega\}$ if position 2 is to be chosen next. Then, the closure function *close* adds the items $f^{\checkmark}\omega\bullet g\omega\omega a$ but avoids the item $f^{\checkmark}\omega\bullet aa$. This is because the former is relevant whereas the latter is not. For the final states 4, 5, 6 and 7, all the unticked symbols are ωs. In state 4, 5 and 6, all variable positions are irrelevant for declaring a match and so all ωs are unticked. However, in state 7, the symbol ω at position 2 is checked because it needs to be labelled with a symbol that is distinct from a and g.

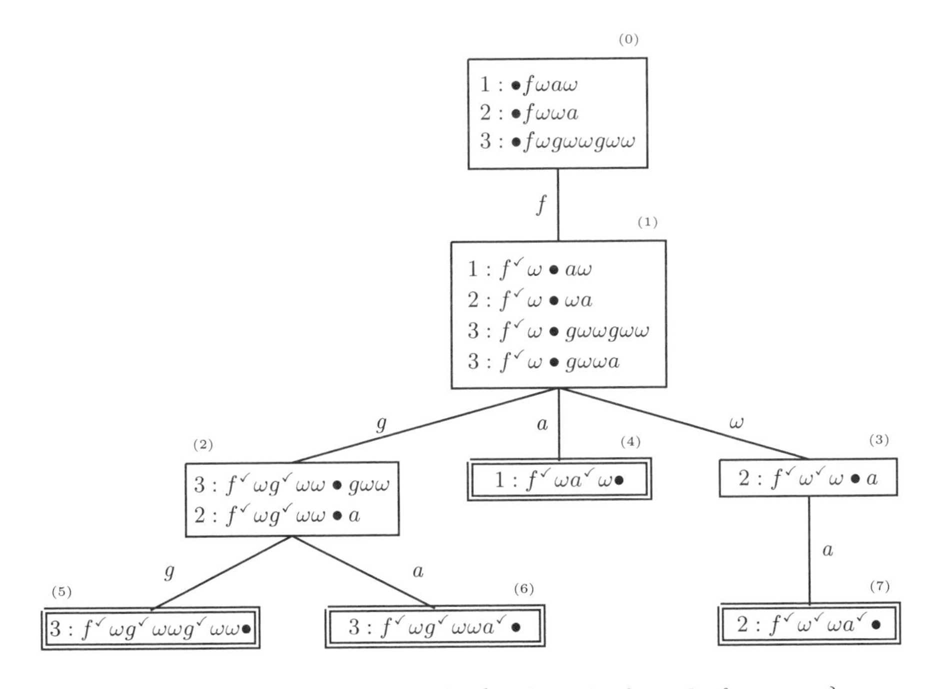

Fig. 4.1. An adaptive automaton for $\{1 : f\omega a\omega, 2 : f\omega\omega a, 3 : f\omega g\omega\omega g\omega\omega\}$

4.4 Heuristics for Good Traversal Orders

In this section, we show how to design selection strategies to choose positions that need to be inspected. This is done using kernels of matching sets as the patterns added by close do not affect the choice. First, we define positions, called *indexes* whose inspection certainly improves matching time and termination of term evaluation. Then, we define *partial indexes*, whose inspection attempts to improve matching times for patterns of high priority. We conclude by prescribing a *good* traversal order for patterns that are not essentially strongly sequential.

4.4.1 Inspecting Indexes First

Knowing that a placeholder can match any term, a context u of a term t is a term containing placeholders such that t is an instance of u. Indexes for a context u with respect to a pattern set Π are placeholder positions in u that must be inspected to announce a match for any pattern in Π that unifies with u. Therefore, in any matching automaton, indexes for a context u must occur in every branch with final state recognising a pattern which is an instance of u.

Definition 10. Let $i = \alpha \bullet s\beta$. A position $p \in \overline{up}(\{i\}, s)$ is an index for i if, and only if, for every term t that matches $\alpha s\beta$, $t[p]$ is a function symbol. For a matching set M and a symbol s, position $p \in \overline{up}(M, s)$ is said to be an *index* for M if, and only if, p is an index for every item in M with matching symbol s.

The normalisation of subterms occurring at indexes is necessary. So, safer adaptive automaton (w.r.t. termination properties) can be obtained by selecting indexes first whenever such positions exist and using one of the heuristics otherwise. Furthermore, adaptive automata that choose indexes first have a smaller (or the same) size than those that do not [11].

4.4.2 Selecting Partial Indexes

Here, we assume, as for most functional languages that deal with overlapping patterns, that the priority rule is a total order for the pattern set. That is, for every pair of patterns π_1 and π_2 in the pattern set, there exists a priority relationship between them. Example of such priority rule include textual priority rule which is used in most functional languages.

For equational programs that are not strongly sequential [3], there are some matching sets with no index. When no index can be identified for a matching set, a heuristic that consists of selecting a partial indexes can be de used. A partial index is an index for a maximal number of consecutive items in a matching set with respect to pattern priorities. Partial indexes can be defined as follows:

Definition 11. Let $M = \{\alpha_1 \bullet s\beta_1, \ldots, \alpha_k \bullet s\beta_k, \ldots, \alpha_n \bullet s\beta_n\}$ be a matching set ordered according to pattern priorities. Then a position $p \in \overline{up}(\{i\}, s)$ is a partial index for M if, and only if, p is an index for $\alpha_1 \bullet s\beta_1, \ldots, \alpha_k \bullet s\beta_k$, p is not an index for $\alpha_{k+1} \bullet s\beta_{k+1}$, and no position p' is an index for $\alpha_1 \bullet s\beta_1, \ldots, \alpha_{k+1} \bullet s\beta_{k+1}$.

Of course, there are other possible ways of defining partial indexes. For example, one may choose to define a partial index for a matching set as the position which is an index for a maximal number of items in the matching set. Another way consists of associating a weight to each pattern and another to each position in the patterns. A pattern weight represents the probability of the pattern being matched whilst a position weight is the sum of the probability of the argument rooted at that position leading to a non-terminating reduction sequence. Let Π be a pattern set and p a position in the patterns of Π. At run-time, the sum of the products of the weight of each pattern and the weight of position p determines the probability of having a non-terminating sequence at p. One may consider to minimise such a sum for partial indexes to improve termination. However, pattern and position weights may be difficult to obtain. Choosing partial indexes defined an in Definition 11 attempts to improve the matching times for terms that match patterns of high priority among overlapping ones. It may also reduce the breadth of the matching automaton as the closure operation adds only few patterns. Except for final matching sets which have no partial indexes, such a position can always be found. The motivation behind choosing such partial indexes is that we believe patterns of high priority are of some interest to the user.

4.4.3 A Good Traversal Order

It is clear that a good traversal order selects indexes first whenever possible. If a matching set has more than one index then the index with a minimal number of distinct symbols must be chosen. Doing so allows us to reduce the breadth of the automaton but does not affect the height of any branch in the subautomaton yielded by the matching set. This because all the positions are indexes and so must occur in every branch of the subautomaton. Assume that p and q are two indexes for a matching set M and labelled in the items of M with n and m distinct symbols respectively where $n < m$. Then inspecting p first would yield n new states and each of which would yield m subautomata which visit the rest of unchecked positions in the items. However, when q is visited first, m new states would be added and each of which would yield n subautomata to inspect the rest of the unchecked positions in the items. Notice that in whichever order p and q are visited, the subautomata yielded after the inspection of p and q are identical. Hence, the size of the automaton that inspects p first has $m - n$ fewer states than that that visits q first. Now, if there is more than one index with a minimal number of distinct symbols then either of them can be selected as the choice affects neither the height nor the breadth of the automaton. When no index can be found for a matching set, a partial index can be selected. If more than one partial index is available then some heuristics can be used to discriminate between them.

A good traversal order should allow a compromise between three aspects. These are *termination*, *code size* and *matching time*. A traversal order that always avoids infinite rewriting sequences when this is possible, satisfies the termination property. However, for non-sequential rewriting systems, this not always possible for all patterns. So a good traversal order for non-sequential systems should maximise this property. The code size of associated with a given traversal order is generally represented by the number of states that are necessary in the corresponding adaptive matching automaton. A good traversal order should minimise the code size criterion. Finally, the last but not least characteristic of a traversal order is the necessary time for a pattern match to be declared. In general, the matching time with a given traversal order is the length or number of visited positions in the longest path of the tree that represent that traversal order. A good traversal order should minimise the matching time.

4.5 Genetically-Programmed Matching Automata

Constructing adaptive pattern matching automata for non-sequential patterns [11] is an *NP*-complete problem. Genetic programming constitutes a viable way to obtain optimal adaptive automata for non-sequential patterns. As far as we know, this is the first work that attempts to apply genetic programming to the pattern matching problem. In the evolutionary process, individuals represent adaptive matching automata for the given pattern set.

Starting form random set of matching automata, which is generally called *initial population*, in this work, genetic programming breeds a population of automata through a series of steps, called *generations*, using the Darwinian principle of natural selection, recombination also called *crossover*, and *mutation*. Individuals are selected based on how much they adhere to the specified constraints, which are termination, code size and matching time. Each traversal is assigned a value, generally called its *fitness*, which mirrors how good it is.

Here, genetic programming [10] proceeds by first, randomly creating an initial population of matching automata; then, iteratively performing a generation, which consists of going through two main steps, as far as the constraints are not met. The first step in a generation assigns for each automaton in the current population a fitness value that measures its adherence to the constraints while the second step creates a new population by applying the two genetic operators, which are crossover and mutation to some selected individuals. Selection is done with on the basis of the individual fitness. The fitter the program is, the more probable it is selected to contribute to the formation of the new generation. Crossover recombines two chosen automata to create two new ones using single-point crossover or two-points crossover as shown in next section. Mutation yields a new individual by changing some randomly chosen states in the selected automaton. The number of states to be mutated is called *mutation degree* and how many individuals should suffer mutation is called *mutation rate.*

4.5.1 Encoding of adaptive matching automata

A state of the adaptive matching automaton can be represented by the corresponding term traversal order. A state is the pair formed by a matching set and its matching position. A simplified representation of adaptive matching automata is that in which the matching set associated with a state is ignored and so concentrates on the matching position. For instance, the matching automaton of Figure 4.1 can be simplified into that of Figure 4.2 in which a state is a pair of a state number and the matching position. Final states are labelled with the number of the rule whose pattern has been matched.

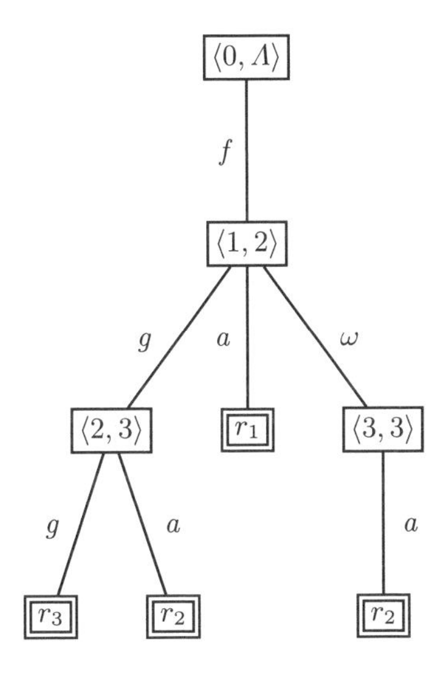

Fig. 4.2. Simplified representation of the adaptive automaton of Figure 4.1 or corresponding traversal order

In order to be able to implement the evolutionary process, we encode the individuals, which represent matching automata, into their traversal orders. Note that for given pattern set, a traversal order completely define the adaptive matching automaton and vice-versa. Note that internally the traversal order of Figure 4.2 is represented as in Figure 4.3. Given the access to a node n, we can access the inspected position by $n[0]$, $n[i]$, for $1 \leq i \leq \#n[0]$ to access the label of the branch i of node n and n_i, for $1 \leq i \leq \#n$ to access to

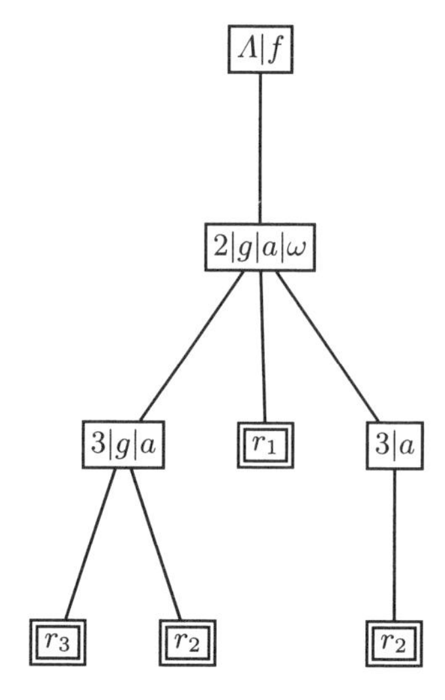

Fig. 4.3. Internal representation of the traversal order of Figure 4.2

access the branches themselves, wherein $\#n$ denotes the number of branches node n has.

4.5.2 Decoding of Traversal Orders

Once the evolutionary process converges to an optimal traversal order for the given pattern set, we can simply construct the corresponding automaton. For instance, reconsider the set of patterns $\Pi = \{r_1 : f\omega a\omega, r_2 : f\omega\omega a, r_3 : f\omega g\omega\omega g\omega\omega\}$. The traversal order of Figure 4.2 can only be decoded into the automaton of Figure 4.1 while the traversal order of Figure 4.4 is decoded into the adaptive matching automaton of Figure 4.5.

4.5.3 Genetic Operators

Crossover recombines two randomly selected individuals into two fresh offspring. It may be *single-point* or *double-point* or *uniform* crossover [12]. Crossover of circuit specification is implemented using single-point and double-point crossovers as described in Fig. 4.6 and Fig. 4.7 respectively.

Mutation acts on a chosen node of the traversal order tree. The chosen node can represent an non-final or a final state in the associated matching automaton. An non-final state can be either mutated to a another non-final state as in Figure 4.8 or to a final state as in Figure 4.9. In the former case the inspected position is replaced by another possible position while in the latter one, a rule number is randomised and used. In case the chose node to be mutated, a rule number different from the one labelling the node is selected and used.

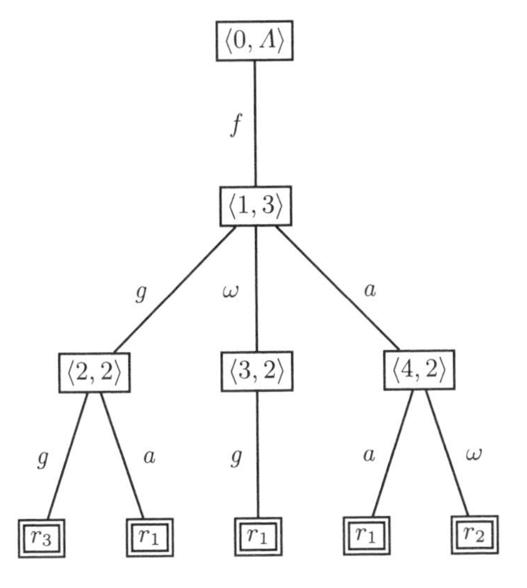

Fig. 4.4. Another possible traversal order for Π

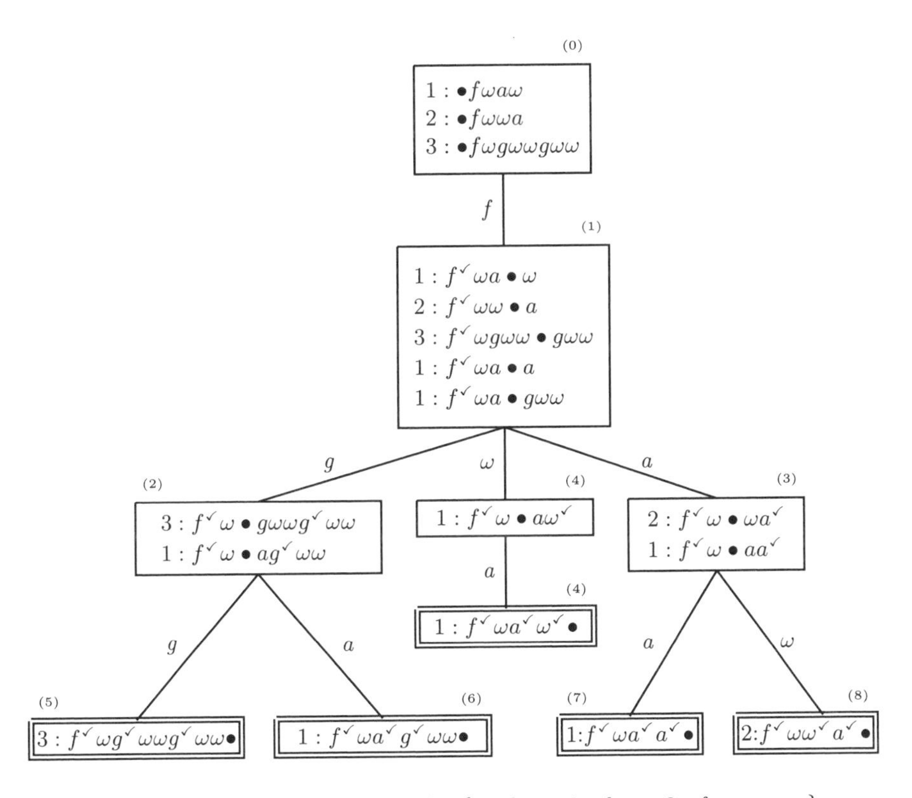

Fig. 4.5. An adaptive automaton for $\{1 : f\omega a\omega, 2 : f\omega\omega a, 3 : f\omega g\omega\omega g\omega\omega\}$

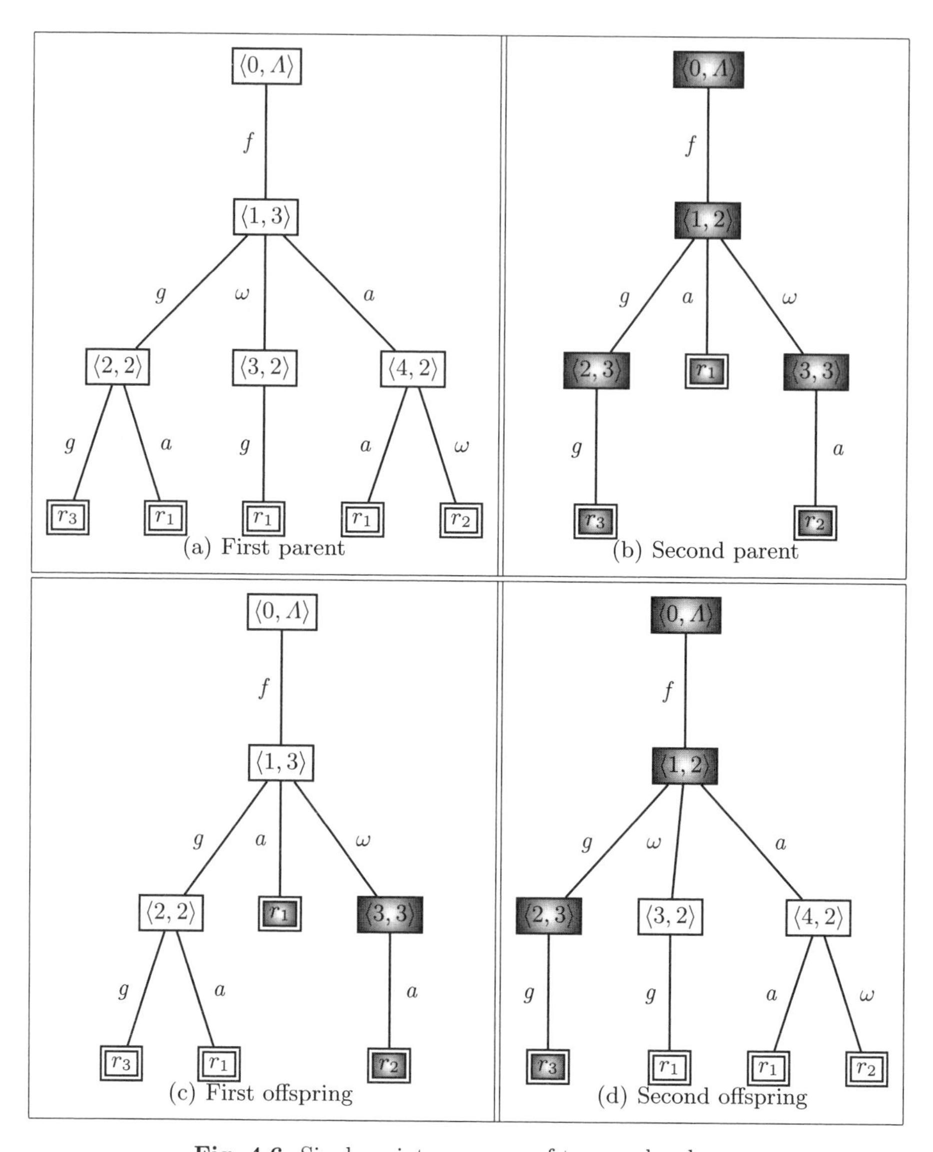

Fig. 4.6. Single-point crossover of traversal orders

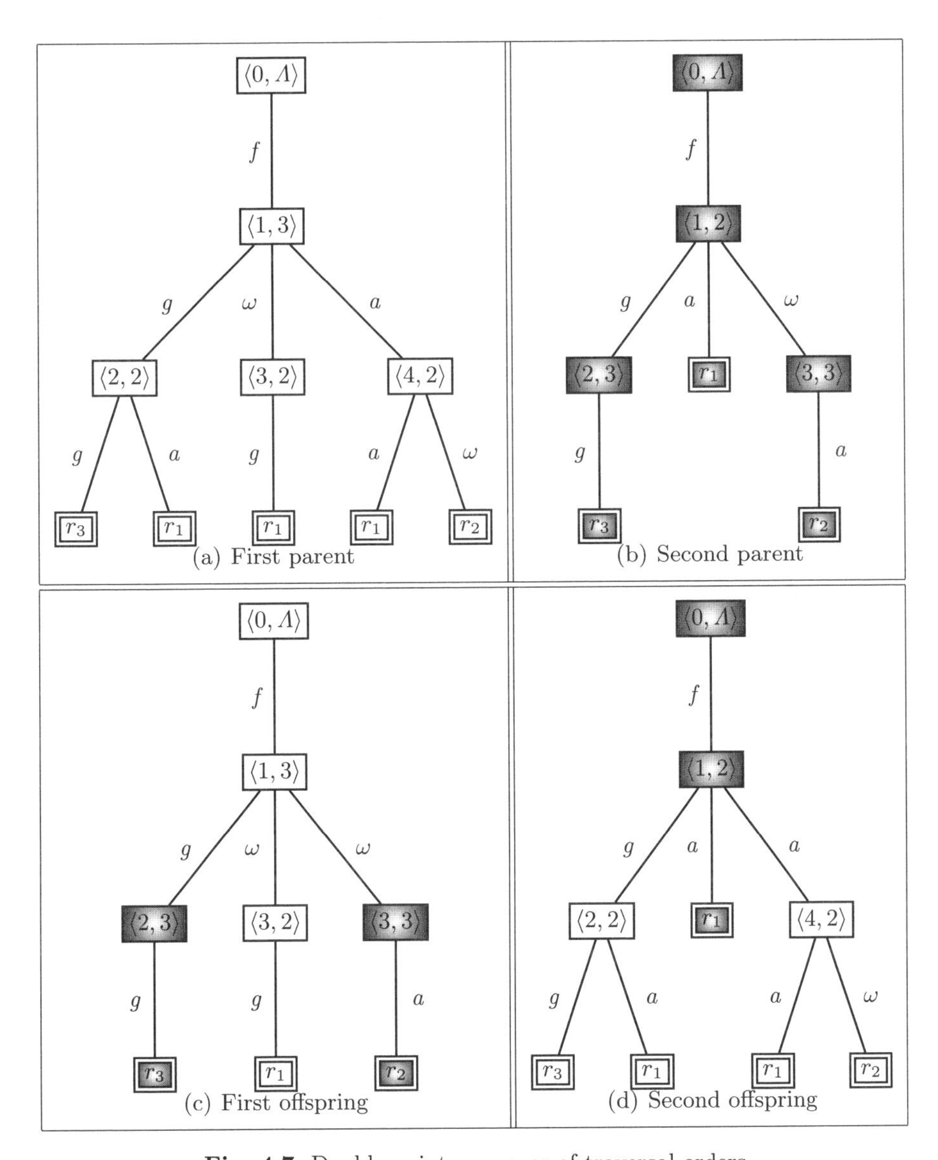

(a) First parent

(b) Second parent

(c) First offspring

(d) Second offspring

Fig. 4.7. Double-point crossover of traversal orders

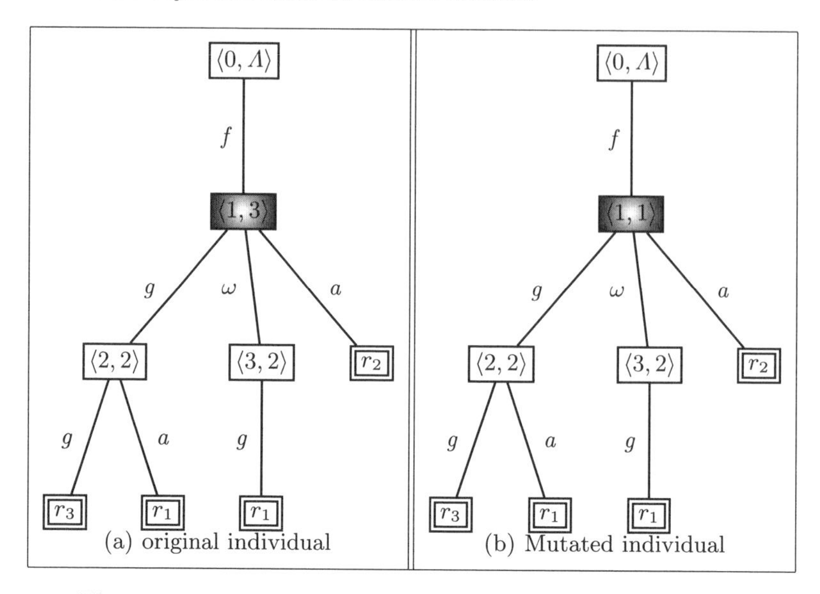

Fig. 4.8. Mutation of non-final state to another non-final state

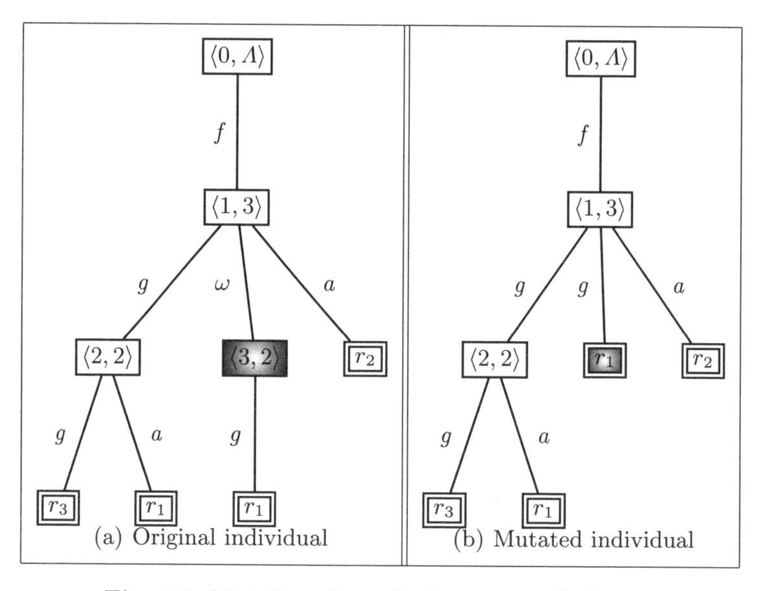

Fig. 4.9. Mutation of non-final state to a final state

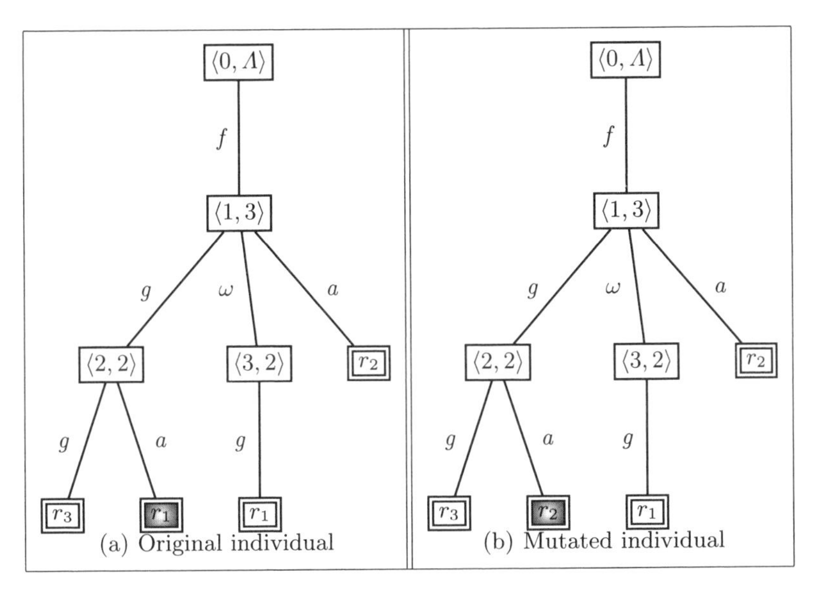

(a) Original individual (b) Mutated individual

Fig. 4.10. Mutation of final state to another final state

4.5.4 Fitness function

When applied to a traversal order τ, the fitness function numerically measures four distinct aspects:

1. how sound is τ with regards to the given pattern set. In other terms, the function evaluates if the inspected positions for each pattern are valid for that pattern. Given a pattern π, an inspected position may not be valid for two reasons: either it does not belong to the set of possible positions for π or its parent-position, which leads to it, was not inspected yet. For instance, let $\pi = fa\omega g\omega\omega$, the set of valid position is $\{\Lambda, 1, 2, 3, 3.1, 3.2\}$. So if, for instance, positions 4 and/or 2.2 are present in the path in τ that declares a match for pattern π then τ is not a correct matching automaton for π. Moreover, if position 3.1 or 3.2 are inspected before or without position 3 being inspected, then τ is also not a correct automaton for π. To measure this soundness factor, which we will denote by $\mathscr{S}$, we sum up a non-soundness penalty $\xi_{\mathscr{S}}$ for each invalid position, and so we can quantify the relation of soundness between two traversal orders τ_1 and τ_2. For pattern set Π, We say that τ_1 *is sounder than* τ_2 if and only if $\mathscr{S}(\tau_1, \Pi) < \mathscr{S}(\tau_2, \Pi)$. The soundness factor for traversal order τ with

respect to pattern set Π is defined as in (4.8). Note that for fully valid traversal orders, the soundness factor is zero. In the definition of (4.8) and for the sake of functional programming, we use an extra parameter, which we call P and represent the set of positions present in the path visited by the function. Initially, this parameter is the empty set.

$$\mathscr{S}(\tau, \Pi, P) = \begin{cases} \sum_{\forall p \in P, p \ni \Pi_{\tau[0]}} \xi_S, & \text{if } \#\tau = 0 \\ \xi_S + \sum_{i=1}^{\#\tau} \mathscr{S}(\tau_i, \Pi, P \cup \tau_i[0]), & \text{if } \tau[0] = x.p \vee \forall y \in P, x \neq y \\ \sum_{i=1}^{\#\tau} \mathscr{S}(\tau_i, \Pi, P \cup \tau_i[0]), & \text{if } \tau[0] = x.p \vee \exists y \in P, x = y \end{cases} \tag{4.8}$$

2. how far does τ satisfy the termination property. In general, termination is compromised when the traversal order inspects positions that are not indexes for the patterns in question. Considering the traversal order tree, this appears in the branches that are labelled by ω. So to quantify the termination factor for τ, which we will denote by $\mathscr{T}(\tau)$, we sum up a non-termination penalty $\xi_{\mathscr{T}}$ for each time ω appears as label in it. The termination factor for traversal order τ is defined as in (4.9). Note that traversal orders that allow terminating rewriting sequences for all patterns have a termination factor zero. However, this is only possible for sequential patterns. Here, we deal with the more complicated case non-sequential patterns.

$$\mathscr{T}(\tau) = \begin{cases} 0 \text{ if} & \#\tau = 0 \\ \xi_T + \sum_{i=1}^{\#\tau} \mathscr{T}(\tau_i), & \text{if } \exists j, 1 \leq i \leq \#\tau | \tau[i] = \omega \\ \sum_{i=1}^{\#tau} \mathscr{T}(\tau_i), & \text{if } \forall i, 1 \leq i \leq \#\tau | \tau[i] \neq \omega \end{cases} \tag{4.9}$$

3. how reduced is the code size of the automaton associated with τ. As state before, the code size of a matching automaton is proportional to the number of its states. So the code size for τ, which we will denote by $\mathscr{C}(\tau)$, is simply the total state number in its traversal order. This defined as in (4.10).

$$\mathscr{C}(\tau) = \begin{cases} 1 & \text{if } \#\tau = 0 \\ 1 + \sum_{i=1}^{\#\tau} \mathscr{C}(\tau_i), & \text{otherwise} \end{cases} \tag{4.10}$$

4. how fast is pattern matching using τ. For a given pattern set, the time spent to declare a match of one the patterns is proportional to the number of position visited. This can be roughly estimated by the length or number of position in the longest path in the traversal order. However, a finer measure of the matching time consists of computing the mean matching time for each pattern in the considered set, which is the average length of all the paths that lead to the match of a given pattern, then computing the mean time considering all the patterns in the set and using the so computed mean matching time for each one of them. The matching time for τ will be denoted by $\mathscr{M}(\tau)$ and is defined in (4.11).

$$\mathscr{M}(\tau) = \begin{cases} 1 & \text{if } \#\tau = 0 \\ 1 + \max_{i=1}^{\#\tau} \mathscr{M}(\tau_i), & \text{otherwise} \end{cases} \tag{4.11}$$

The evolution of adaptive matching automata requires a multi-objective optimisation. There are several approaches. Here, we use an aggregation selection-based method to solve multi-objective optimisation problems. It aggregates the multiple objectives linearly into a single objective function. For this purpose, it uses weights to compute a weighted sum of the objective functions. The optimisation problem of adaptive matching automata has 4 objectives, which are the minimisation of *soundness* ($\mathscr{S}$), *termination* ($\mathscr{T}$), *code size* ($\mathscr{C}$) and *matching time* ($\mathscr{M}$). The is transformed into a single objective optimisation problem whose objective function is as (4.12):

$$\min_{\forall \tau} \mathscr{F}(\tau, \Pi) = \Omega_1 \mathscr{S}(\tau, \Pi) + \Omega_2 \mathscr{T}(\tau) + \Omega_3 \mathscr{C}(\tau) + \Omega_4 \mathscr{M}(\tau) \tag{4.12}$$

wherein Ω_i are non-zero fractional numbers, usually called *weights* or *importance factors*, such that we have $\Omega_1 + \Omega_2 + \Omega_3 + \Omega_4 = 1$.

Note the a weight value Ω_i defines the importance of objective i. Therefore, to put more emphasis on valid adaptive matching automata, we give more importance to the soundness objective. The remaining objectives are of equal importance.

4.6 Comparative Results

Optimisation using the weighted sum approach is extremely simple. It consists of choosing a vector of weights and performing the optimisation process as a single objective problem. As a single or limited number of solutions, hopefully Pareto-optimal [19], is expected from such process, it is necessary to perform as many optimisation processes as the number of required solutions [16]. In our experiments, we used the [0.4, 0.2, 0.2, 0.2] as weight vector.

The pattern matching automata for some known problems were used as benchmarks to assess the improvement of the evolutionary matching automata with respect to the ones designed using known heuristics. These problems were first used by Christian [2] to evaluate his system HIPER. The *Kbl* benchmark is the ordinary three-axiom group completion problem. The *Comm* benchmark is the commutator theorem for groups. The *Ring* problem is to show that if $x^2 = x$ is a ring then the ring is commutative. Finally, the *Groupl* problem is to derive a complete set of reductions for Highman's single-law axiomatisation of groups using division.

For each of the benchmarks, we built the both matching automata, i.e. using genetic programming and using known heuristics and obtained the number of necessary states, matching time required. This should provide an idea about the size of the automaton. Furthermore, we obtained the evaluation times of a given subject term under our rewriting machine (for details see [14]) using both matching automata as well as the evaluation times of the same subject terms under the system HIPER. The space and time requirements are given in Table 4.1.

Table 4.1. Space and time requirements for miscellaneous benchmarks

Benchmark	Number of states Classic	Number of states Evolutionary	Time (s) Classic	Time (s) Evolutionary
Kbl	49	28	0.079	0.002
Comm	153	89	1.229	0.211
Ring	256	122	2.060	0.565
Groupl	487	205	1.880	0.704

4.7 Summary

In this chapter, we presented a novel approach to generate adaptive matching automata for non-sequential pattern set using genetic programming. we first defined some notation and necessary terminologies. Then, we formulated the problem of pattern matching and the impact that the traversal order of the patterns has on the process efficiency, when the patterns are ambiguous. We also gave some heuristics that allow the engineering of a relatively good traversal order. In the main part of the chapter, we described the evolutionary approach that permits the discovery of traversal orders using genetic programming for a given pattern set. For this purpose, we presented how the encoding of traversal orders is done and consequently how the decoding of an evolved traversal order into the corresponding adaptive pattern-matcher. We also developed the necessary genetic operators and showed how the fitness of evolved traversal orders is computed. We evaluated how sound is the obtained traversal. The optimisation was based on three main characteristics for matching automata, which are termination, code size and required matching time. Finally, we compared evolutionary adaptive matching automata, obtained for some universal benchmarks, to their counterparts that were designed using classic heuristics.

References

1. A. Augustsson, A Compiler for Lazy ML, Proc. ACM Conference on Lisp and Functional Programming, ACM, pp. 218-227, 1984.
2. J. Christian, Flatterms, Discrimination Nets and Fast Term Rewriting, Journal of Automated Reasoning, vol. 10, pp. 95-113, 1993.
3. D. Cooper and N. Wogrin, Rule-Based Programming with OPS5, Morgan Kaufmann, San Francisco, 1988.
4. N. Dershowitz and J.P. Jouannaud, Rewrite Systems, Handbook of Theoretical Computer Science, vol. 2, chap. 6, Elsevier Science Publishers, 1990.
5. A.J. Field and P.G. Harrison, Functional Programming, International Computer Science Series, 1988.
6. J.A Goguen and T. Winkler, Introducing OBJ3, Technical report SRI-CSL-88-9, Computer Science Laboratory, SRI International, 1998.
7. A. Gräf, Left-to-Right Tree Pattern-Matching, Proc. Conference on Rewriting Techniques and Applications, Lecture Notes in Computer Science, vol. 488, pp. 323-334, 1991.
8. C.M. Hoffman and M.J. O'Donnell, Pattern-Matching in Trees, Journal of ACM, vol. 29, n. 1, pp. 68-95, 1982.
9. P. Hudak and al., Report on the Programming Language Haskell: a Non-Strict, Purely Functional Language, Sigplan Notices, Section S, May 1992.
10. J.R. Koza, Genetic Programming. MIT Press, 1992.
11. A. Laville, Comparison of Priority Rules in Pattern Matching and Term Rewriting, Journal of Symbolic Computation, n. 11, pp. 321-347, 1991.
12. J.F. Miller, P. Thompson and T.C. Fogarty, Designing Electronics Circuits Using Evolutionary Algorithms. Arithmetic Circuits: A Case Study, In Genetic Algorithms and Evolution Strategies in Engineering and Computer Science, Quagliarella et al. (eds.), Wiley Publisher, 1997.
13. N. Nedjah, C.D. Walter and S.E. Eldridge, Optimal Left-to-Right Pattern-Matching Automata, Proc. Conference on Algebraic and Logic Programming, Southampton, UK, Lecture Notes in Computer Science, Springer-Verlag, vol. 1298, pp. 273-285, 1997.
14. N. Nedjah, C.D. Walter and S.E. Eldridge, Efficient Automata-Driven Pattern-Matching for Equational programs, Software-Practice and Experience, John Wiley Eds., vol. 29, n. 9, pp. 793-813, 1999.
15. N. Nedjah and L.M. Mourelle, Implementation of Term Rewriting-Based Programming Languages, Hauppauge, NY, ISBN 1594-5439-09, 2005.
16. N. Nedjah and L.M. Mourelle (Eds.), Real-World Multi-Objective Systems Engineering, Nova Science Publishers, Hauppauge, NY, ISBN 1590-3364-53, 2003.
17. N. Nedjah and L.M. Mourelle, More Efficient Left-to-Right Matching for Overlapping Patterns, vol. 3, n. 2-4, pp. 230–247, 2005.
18. M.J. O'Donnell, Equational Logic as Programming Language, MIT Press, 1985.

19. V. Pareto, Cours d'économie politique, volume I and II, F. Rouge, Lausanne, 1896.
20. R.C. Sekar, R. Ramesh and I.V. Ramakrishnan, Adaptive Pattern-Matching, SIAM Journal, vol. 24, n. 5, pp. 1207-1234, 1995.
21. D.A. Turner, Miranda: a Non Strict Functional Language with Polymorphic Types, Proc. Conference on Lisp and Functional Languages, ACM, pp. 1-16, 1985.
22. P. Wadler, Efficient Compilation of Pattern-Matching, In "The Implementation of Functional Programming Languages", S. L. Peyton-Jones, Prentice-Hall International, pp. 78-103, 1987.

5

Genetic Programming in Data Modelling

Halina Kwasnicka[1] and Ewa Szpunar-Huk[2]

[1] Institute of Applied Informatics, Wroclaw University of Technology, Wyb. Wyspianskiego 27, 50-370 Wroclaw, Poland halina.kwasnicka@pwr.wroc.pl
[2] Institute of Applied Informatics, Wroclaw University of Technology, Wyb. Wyspianskiego 27, 50-370 Wroclaw, Poland ewa.szpunar-huk@pwr.wroc.pl

The chapter presents some abilities of Genetic Programming (GP), one of the branches of Evolutionary Computation (EC), which becomes very popular in different areas, such as Data Modelling (DM). In this chapter attention is focused on using GP to make data collected in large databases more useful and understandable. Here we concentrate on such problems as mathematical modelling, classification, prediction and modelling of time series. Successive sections are dedicated to solving the above mentioned problems using the GP paradigm. Sections devoted to individual problems share a similar structure: their first parts characterise the specificity of a given problem with respect to applying GP to the task under consideration, the second parts describe examples of using GP in such a problem and the final parts contain general information connected with the issue being discussed in the first two parts. The chapter begins with a short introduction to Data Modelling and Genetic Programming. The last section recapitulates the chapter.

5.1 Introduction

Evolutionary Computation paradigm becomes very popular. A number of approaches that exploit the biological evolution ability to adaptation arise. Between them we can distinguish Genetic Algorithms, Evolutionary Strategies and Evolutionary Programming. Genetic Programming (GP) proposed by Koza use the same idea as Genetic Algorithm (GA) and it is perceived as a part of GA. The main difference is that GP use different coding of potential solutions, what forces different genetic operators such as crossover and mutation. The main idea of Evolutionary Computation is to imitate natural selection, that is the mechanism that relates chromosomes with the efficiency of the entity they represent, thus allowing those efficient organisms, which are

H. Kwasnicka and E. Szpunar-Huk: *Genetic Programming in Data Modelling*, Studies in Computational Intelligence (SCI) **13**, 105–130 (2006)
www.springerlink.com

well-adapted to the environment, to reproduce more often than those which are not.

Genetic Programming is strongly developed in different places – universities, laboratories etc. The web http://www.genetic-programming.org/ is a reach source of knowledge dedicated to GP. The readers can find there useful both theoretical and practical information. Genetic Programming is very popular everywhere, where the algorithm to solve a considered problem is not known, because GP starts from a high-level statement of "what needs to be done". A computer program solving the problem is evolved, what means – it is developed automatically.

The chapter concerns the problem connected with making a great amount of data, collected in different places (firms, consortia, etc.), useful. Discovering knowledge from possessed data and making it useful is very important especially for managers.

The chapter, beside Introduction and Summary, consists of the three main sections. Section 5.2 gives introduction and the example of using Genetic Programming for mathematical modelling task. Section 5.3 details the Genetic Programming used for classification task, where of course, we can use some rules induction and decision trees generation algorithms, but as we can see, in same cases Genetic Programming can reveal some advantages. Other important problems are connected with prediction and time series modelling. GP can also be used for these tasks. The third main section (Section 5.4) presents how GP can be used for such problems. Some conclusions are placed in the last section – Summary (Section 5.5).

5.2 Genetic Programming in Mathematical Modelling

Mathematical modelling can be regarded as a search for equations that provide a good fit to numerical data. It is used in variety of problems like scientific law discovery or prediction tasks in fields where the theory does not give definite answers about the outcome of experiments. The search process however using traditional methods, for example polynomial regression, is often difficult, because in these methods a specific model form must be assumed, what demands strong theoretical knowledge. Usually in case of regression only the complexity of this model can be varied so the search task is often limited to finding a set of coefficients. GP allows searching for a good model in a different, more "intelligent" way, and can be used to solve highly complex, non-linear, probably chaotic or currently not fully or properly understood problems [1].

Induction of mathematical expressions on data using GP is introduced by Koza and is called symbolic regression [2], to emphasise the fact that the object of search is a symbolic description of a model, not just a set of coefficients in a prespecified model. Koza showed that GP could perform well in function identification tasks, because it searches the space of all possible models for both a good model form and the appropriate numeric coefficients for the model. The search space is constrained only by the available model pieces (variables, functions, binary operators, syntax rules). Furthermore, GP is not obligated to include all input variables in the equation, so it can perform a kind of dimensionality reduction.

5.2.1 Adaptation of GP to Mathematical Modelling

Genetic programming can be successfully used to find a mathematical data model because it can find the shape of the equation as well as the composition of primitive functions, input variables and values of coefficients at the same time, without any additional information except the performance of the analysed expression on the data. But the key issue to make GP algorithm work is a proper representation of a problem, because GP algorithms directly manipulate the coded representation of equation, which can severely limit the space of possible models. An especially convenient way to create and manipulate the compositions of functions and terminals are the symbolic expressions (S-expressions) of the LISP programming language. In such a language the mathematical notation is not written in standard way, but in prefix notation. It allows building expression trees out of these strings, that can be easily evaluated then. Some examples of expressions written in algebraic and prefix notation with corresponding trees are shown in Fig. 5.1.

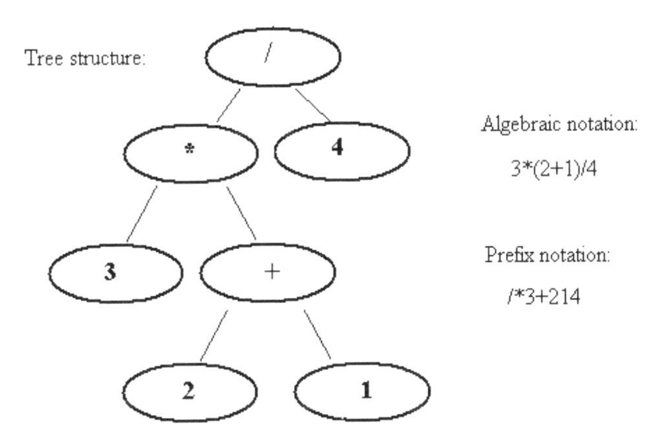

Fig. 5.1. An examplary tree in GP and the coded expression

In the Genetic Programming paradigm, the individuals in the population are compositions of functions and terminals appropriate to the particular problem's domain. Functions and terminals are adequately nodes and leaves of the trees being evolved by the GP. Therefore, before running GP to search for an equation, it is necessary to define operators dedicated to the considered task. For the mathematical modelling problem the four binary arithmetic functions (+, -, *, /) are usually used, but other specific functions (e.g., trigonometric functions, sqrt, abs) can also be useful. The terminals or leaves are either variables or coefficients. Variables correspond to the data being analysed, while coefficients are defined by numerical ranges.

Using arithmetic functions as operators in GP has some drawbacks. One of them is that the closure property of GP requires, that each of the base functions is able to accept, as its arguments, any value that might possibly be returned by any base function and any value that may possibly be assumed by any terminal. This condition is not satisfied by e.g., division operator. That is why a protected division function $div(a, b)$ is usually defined which returns 1 in two cases: when $a = b$ and when $b = 0$. Otherwise it returns a result of division a/b. Similar protections must be established for many other functions as e.g., $log(a)$ or $sqrt(a)$. The problem with closure property is entirely eliminated in Strongly Type Genetic Programming (STGP) [4], in which it is assumed that each function specifies precisely the data types of its arguments and its returned values, but on the other hand, additional modifications of the searching process are required, like creating more individuals in each generation in order to better explore the search space.

During the evaluation of GP solution some numerical problems may also easily occur, like underflows or overflows caused by extremely small or high values or major precision loss, e.g., determining a trigonometric function such as $sin(a)$ for some value a much larger than 2Π [5]. It is also assumed that the creation of floating-point constants is necessary to do symbolic regression and it seems to be a weak point, because coefficient must be assembled from random values, so destination functions containing constants are usually much more demanding to GP than others. Designing GP algorithm to mathematical modelling it is necessary to consider most of these problems.

5.2.2 An educational Example – the Hybrid of Genetic Programming and Genetic Algorithm

In this section we present a detailed description of a simple GP algorithm to mathematical model induction. The work was done as a part of wider study of Szpunar-Huk, described in [24]. The goal of the designed algorithm is to find a function $y = f(x_1, ..., x_n)$, which reflects potentially existing relationship between input and output values. As the input data we have a sample set $S = \{x_1, ..., x_n, y\}$, where $x_1, ..., x_n$ are independent input values and y is an output value (assumed function of x_i). Values $x_1, ..., x_n$ and y are continuous numbers converted to floating point numbers with a specified for an individual experiment precision. The algorithm is tested to examine its ability to do symbolic regression on sample sets of different size and level of noise, generated using predefined mathematical functions, and for different evolution control parameters (e.g., population size, probabilities of crossover and mutation).

During experiments generations' numbers with standard deviations needed to find solutions and obtained errors were analysed.

Firstly it is important to define terminals and functions sets used for approximating the test functions. Let us assume:

- The terminal set $T = \{x_1, ..., x_n, R\}$ – the set of possible values of leaves of trees being evolved contains input variables' labels and continuous constants between 1.0 and 100.0.
- The set of available base functions $F = \{+, -, *, /\}$ include just addition (+), subtraction (–), multiplication (*) and division operator (/).

It is important to notice, that in this case the division operator is not protected, so the closure property of GP is not satisfied. This problem is solved by a proper definition of the fitness function. All the base functions have two arguments what significantly simplifies crossover operation.

A fitness function measures how well each individual (expression) in the population performs the considered problem (task). In described experiments two fitness functions were used. The first one was the sum of errors made by an individual on a sample set and could be written as:

$$F = \sum_{i=1}^{N} |ex(x^i) - y^i| \tag{5.1}$$

where N is a number of samples in the set S, $ex(x^i)$ is a value returned by an expression ex (encoded expression by the considered individual) for the ith sample, y^i is a value of output y for i^{th} sample.

The second function was the average percentage errors made by an individual on a sample set and could be written as:

$$F = \frac{1}{N} * \sum_{i=1}^{N} \frac{100 * |ex(x^i) - y^i|}{ex(x^i)} \tag{5.2}$$

where as before, N is a number of samples in the set S, $ex(x^i)$ is a value returned by an expression ex for the ith sample, y^i is a value of output y for i^{th} sample.

For both functions, the closer the sums are to zero, the better the expressions (solutions) are. Additionally, when during evaluation of a given expression on any sample from a sample set, division by zero occurs, the fitness function for the individual is set to "undefined". It is also worth mention, that in case of the first function, when the values returned by the sample are close to zero, it is difficult to distinguish between good and bad individuals, what can significantly affect GP performance, but due to the specific of chosen data it was not important in the described experiments.

In the selection phase couples of individuals for crossover are chosen. As a selection method here is proposed the method based on elite selection. Firstly all individuals are ranking according to their fitness (expressions with "undefined" value are placed on the end of the list). Secondly a number of the best different individuals (the exact number is set by an appropriate parameter – usually it is about 10% of population size) are taken to the next generation without any changes. A reason for this is that in GP during crossover phase expressions cannot evolve well because they are frequently destroyed and it is easy to loose the best ones, what can result in poor performance. Selected individuals must be also different to avoid premature convergence of the algorithm. The remaining couples for the crossover phase are generated as follows: the first individual is the next individual from the ranking list (starting from the best one), the second one is randomly chosen from the whole population, not excluding expressions with "undefined" value.

After selection a number of couples (according to a given probability parameter) are copied to the next generation unchanged, the rest individuals undergo the crossover – a random point is chosen in each individual and these nodes (along with their entire sub-trees) are swapped with each other, creating two entirely new individuals, which are placed in the new generation. In the classical GP, each tree could be of potentially any size or shape, but from the implementation point of view, every new tree can not be bigger then the predefined maximum depth, so if too deep tree is generated during crossover operation, it is being cut to the maximum allowed size. Here, in the recombination operation another parameter is also introduced. The parameter specifies probability of choosing a leaf as a place of crossing, because choosing a random point can result in fast shortening of all individuals in the population.

Mutation is an additional included genetic operator. During mutation operation nodes or leafs are randomly changed with a very little probability, but over a constraint, that the type of node remains unchanged: a terminal is replaced by an another terminal and a function by an another function. It is difficult to estimate the real size of a population mutation since, during reproduction phase – when maximum depth of tree is established – trees are often cut, what also can be viewed as a kind of mutation.

An initial population is randomly generated, but creating a tree with maximum depth is more likely to come than creating a tree with only one or just few nodes.

Experiments

To validate described method, some numerical experiments were conducted. In this case some sample sets were generated according to a given function f, the algorithm was looking for this function, but a sample could contain more variables than it was used by the given function. The algorithm turned out to be a good tool to exactly discover plenty of functions. Table 5.1 shows some examples of mathematical expressions, the algorithm managed to find in generated data, and the average number of needed generations of the algorithm. During this experiments the number of input variables' labels in the

Table 5.1. Sample of test functions and number of generations needed to find that function by the algorithm

Function	Number of generations
$y = x_1 x_1 x_1$	3
$y = x_3 + x_2 x_1 - x_4$	8
$y = x_1/(x_2 + x_3)$	16
$y = x_1 + x_1 x_1 - x_2/x_3$	22
$y = x_1 + x_2 + x_3 + x_4 + x_5 + x_8 + x_7 + x_8$	31
$y = x_1 + x_2/x_3 - x_3/x_1$	45
$y = x_3 x_2/(x_5 + x_4) - x_4$	103
$y = x_1 + x_2/x_3 + x_3 x_4 - (x_3 + x_3)/x_2$	135
$y = 10x_1 + 5x_2$	137
$y = x_1 + x_2 + x_3 x_3 - x_1 x_1 x_1/(x_2 x_2)$	207
$y = x_1 + x_2 + x_3 + x_4 + x_5 + x_8 + x_7 + x_8 + x_2 x_3 x_4$	375
$y = x_1 x_1 x_1 + x_2 x_2 x_2 + x_3 x_3 x_3 + x_4 x_4 x_4$	579

sample set and the number of different input variables in the equation where equal. In a number of experiments the vector of input variables x_i contains a variables that are not taken into account during output calculation, they are additional variables. It means that our algorithm looks for the function $y = f(x_1, ..., x_{10})$ but the proper dependency is $y = f(x_1, x_2)$ – only two input variables influence the output variable. In such cases the algorithm could also find the relationship, but in longer time. A number of redundant variables we denote by N_{add}, $T_{solution}$ means an average number of generations for finding solution, but δ_{add} – a standard deviation of $T_{solution}$. Exemplary results are shown in table 5.2. It is important to mention that the algorithm requires

Table 5.2. The result given by GP+GA for different number of additional variables

N_{add}	$T_{solution}$	δ_{add}
0	8.71	1.9
1	17.4	5.91
2	19.7	5.08
3	20.7	6.75
4	29.7	12.4
5	37.4	14.2
6	42.1	30.6

various control parameters to proper working. Some of them are mentioned above. The most important of them with their presumed values are listed below. They are suggested values, often used in the experiments, but in some tasks their values must be carefully tuned.
A list of control parameters of GP used in the experiments:

- Population size: 300-400
- Maximum depth of the tree: 6-8
- Percent of individuals moved unchanged to the next generation: 10-30%
- Crossover probability: 0.85
- Probability of choosing leaf as a crossing point: 0..1
- Mutation probability: 0.01.
- Number of samples: 100-1000

Tuning of the Algorithm

The algorithm in such a form is able to find hidden model in a sample set even with noisy data – the output was generated not exactly from a given function, but outputs were in 5%, 10% or even more noisy. Difficulties however appeared when a tested function included coefficients, especially with real valued coefficients. It is not possible to find relationships, even if it seems to be easy (e.g., $y = 15x_1 + 4x_2 - 23.4x_3/x_2$). Enlarging the initial population size from 500 to 5000 individuals can partially solve this problem, but for plenty of functions it is not sufficient.

To make the algorithm more effective an additional algorithm based on the classic Genetic Algorithm is included. The aim of this algorithm is to find, for a given set of best individuals, after a given number of generations in GP, as good coefficients as possible. That is why the algorithm is cold Tuning Algorithm (TA).

The tuning algorithm works as follows: for the given individual taken from the GP algorithm a number of coefficients is established. This number defines a length of individuals in the TA, where an individual is a sequence of numbers from the given range and with a given precision. An initial population of the TA contains sequences generated on an expression taken from the GP algorithm and random sequences – one hundred individuals altogether. Generated in such a way population is evolved in the standard way: individuals are selected, crossed over and mutated. The fitness function and the selection operation are the same as in the GP algorithm, but the crossover operation is much simpler: for a given couple of individuals a random point in constants sequence is chosen (the same for both individuals) and a part of one individual is swapped with a part of the second one, what is shown in Fig. 5.2. Mutation however is much more important here than in GP algorithm. Its aim is to make it possible to tune the coefficients. A special kind of mutation is proposed. Each real number is a set of Arabic numeral, e.g., 13.86 consists of digits 1, 3, 8 and 6. The probabilities of mutation of particular digits are different – the lowest is for the digit placed on the left (the most meaningful) and systematically is increased for the next digits (going to the right). The scheme of the whole algorithm (GP connected with TA) is shown in Fig. 5.3. The tuning process is performed every N generations of GP for S different, best individuals from an actual generation of the GP. Incorporating of a GA allows approximation of relationships previously impossible to find. For exam-

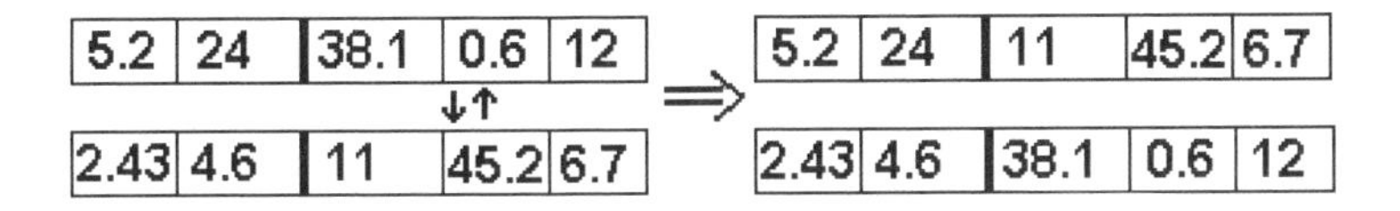

Fig. 5.2. An example of crossover in TA

ple, when the dataset was generated using equation 5.3, the GP has problem with finding the exact coefficients.

$$y = 11.3/x_2 + 10x_3x_3 - 5.2x_2/(x_3 + x_5) + 33.8x_3x_3x_3x_5x_6x_6 + \\ +20(x_4 + x_2)/x_5 + 8x_5x_5x_5 + 12(x_3 + x_5)/x_2 + x_3x_3x_3 \tag{5.3}$$

Using the tuning algorithm worked every 40 generations of the GP on 15 best individuals after 100 generation of GP, the average percentage error of the best individual on dataset was equal to 1.2% while without tuning only 9.25%.

Input: X, Y
Output: Expression $y' \approx f(\mathbf{X})$
1. Create an initial GP population and set GN (number of generations) to 0
2. Evaluate the fitness of the population members
3. While termination criteria not met do
4. If $GN mod N = 0$ perform tuning
5. Take S individuals for tuning
6. For each of S chosen individuals do
7. Create an initial GA population
8. Evaluate the fitness of the population members
9. While termination criteria not met do
10. Probabilistically apply crossover operator
11. Probabilistically apply mutation operator
12. Go to step 8
13. Probabilistically apply crossover operator
14. Probabilistically apply mutation operator
15. Increase GN
16. Go to step 2
17. End.

Fig. 5.3. The scheme of the GP in conjunction with GA used as TA algorithms

5.2.3 General Remarks

In the presented example abilities of GP algorithm to discover mathematical model on artificially generated data where shown, but approximation a function using a given finite sample set of values of independent variables and associated values of dependent variables is a very important practical problem. In practice, the observed data may be noisy and there may be no known way to express the relationships involved in a precise way. That is why GP is widely used to solve the problem of discovering empirical relationships from real observed data. To show, how wide variety of domains GP is successfully applied some of experiments with GP on real data are mentioned below.

Koza [3] showed, how the Genetic Programming paradigm can be used to create an econometric model by rediscovering the well-known non-linear econometric "exchange equation" $M = PQ/V$, which relates the money supply M, price level P, gross national product Q, and the velocity of money V in economy. He also showed how Kepler's Third Law can be rediscovered. Sugimoto, Kikuchi and Tomita were able, by applying some improvements to GP, to predict an equation from time course data without any knowledge concerning the equation. They predicted biochemical reactions and the relative square error of predicted and given time-course data were decreased, due to GP, from 25.4% to 0.744% [6]. Langdon and Barrett have used Genetic Programming (GP) to automatically create interpretable predictive models of a small number of very complex biological interactions, which are of great interest to medicinal and computational chemists, who search for new drug treatments. Particularly, they have found a simple predictive mathematical model of human oral bioavailability [7]. Makarov and Metiu use GP to the analytic solutions of the time-independent Schrodinger equation. They tested their method for a one-dimensional enharmonic well, a double well, and a two-dimensional enharmonic oscillator [8]. Diplock, Openshaw i Turton used GP as a tool for creating new models of complex geographical system. They run parallel GP algorithm on Cray T3D 512 supercomputer to create new types of well performing mathematical model [9].

It is important to notice that in scientific discovery however, obtaining some adequate fit to data is not enough. To fully incorporate some results in the body of scientific work, it is necessary that some form of understanding about the expressions that are induced is achieved. The expressions thus need some further justification before they can be used as models of the phenomenon under study. The resulting equations can then be tested with statistical methods to examine their ability to predict the phenomenon on unseen data.

5.3 Decision Models for Classification Tasks

Classification task is the one of fundamental and the most studied tasks in the area of data mining. Given a set of predetermined, disjoint target classes $C_1, C_2, ..., C_n$, a set of input attributes $A_1, A_2, ..., A_m$, and a set of training data S with each instance associated with a unique target class label, the objective of this task is to find mutual properties among a set of objects in data set which can be used to determine the target category for new unseen data given its input attributes' values. In the other words, this process finds a kind of decision models – a set of classification rules (or procedures in some cases). Classification rules are commonly expressed in the form of IF-THEN rules, where the IF part of a rule contains a logical combination of conditions on the attributes and the THEN part points to a class label. The popular decision model is also a decision tree, in which the leaves are the class labels while the internal nodes contain attribute-based tests and they have one branch emanating for each possible outcome of the test.

Most rules induction (e.g., CN2) and decision trees generation algorithms (e.g., CART and C4.5) perform local, greedy search to generate classification rules, that are relatively accurate and understandable, but often more complex than necessary. There are several properties of GP, which make them more convenient for application in data mining tasks comparing to other techniques. The reason is that the local, greedy search selects only one attribute at a time, and therefore, the feature space is approximated by a set of hypercubes. In real-world applications, the feature space is often very complex and a large set of such hypercubes might be needed to approximate the class boundaries. GP algorithms perform global search. That is why it copes well with attributes interaction problem, by manipulating and applying genetic operators on the functions. Additionally some data mining algorithms work only with qualitative attributes, so numerical values must be divided into categories. GP can deal with any combination of attribute types. Moreover, GP algorithms have ability to work on large and "‘noisy" datasets. They have the high degree of autonomy that makes it possible to discover of knowledge previously unknown by the user.

5.3.1 Adaptation of GP to Clasification Task

Classification task is similar to the regression, and regression could also be used to solve some classification problems but generally, building decision models emphasis on discovering high-level, comprehensible classification rules rather than just producing a numeral value by a GP tree. Application of GP in this field is similar, but due to some differences, causes additional problems.

Problems with Closure Property

The main difference between using GP in classification and regression tasks is that in classification the target class (the goal attribute) is nominal while in regression – continuos one. When data set contains mixture of real-valued and nominal attribute values, different attributes are associated with different functions, so the closure property is usually not satisfied by the standard GP algorithm and the closure problem is much more complicated. One idea to solve the problem is to use logical operators for all inner nodes, while leaves have only boolean values or are represented as functions that return Boolean values. Other approaches to go around the closure requirements are STGP, mentioned in section 5.2.1 [4, 10] (sometimes called constrained-syntax GP), and its variant called Grammar-Based Genetic Programming (GBGP) [10]. The key idea of STGP is such, that for each function f from a set of functions, the user specifies which terminals/functions can be used as a children node containing function f. In GBGP syntactic and semantic constraints are specified by a grammar [10, 14], so the important preparation step for the GP is to identify a suitable target language, in which to evolve programs. It should be underlined, that the success of evolution is dependent upon the use of a language, that can adequately represent a solution to the problem being solved.

Michigan – Pittsburgh Approaches

The next step designing a GP for discovering decision model, after choosing a proper language, is to choose individual representation. There are two main approaches to use GP algorithms to classification rule discovery, based on how rules are encoded in the population of individuals: the Michigan [11] and the Pittsburgh approach [11]. In the Michigan approach each individual encodes a single classification rule, whereas in the Pittsburgh approach each individual encodes a set of classification rules. The choice of individual representation determines the run of GP algorithm. The Michigan is best suitable for binary classification problem, where there are only two classes to distinguish and one individual is entire solution for the problem. The problem gets more complicated when there are more then two classes in the classification task but this approach is also used in such problems. The most common solution for this in the literature is to run the GP k times, where k is the number of classes [12]. In the ith $(i = 1, ..., k)$ run, the GP discovers rules for the ith class. When GP is searching for rules for a given class, all other classes are merged into a large class. In the Pittsburgh approach GP can work with a population of individuals where each individual corresponds to a rules set, constituting a complete solution for the classification problem, i.e. entire decision tree [11].

Fitness Function Defining

Discovering useful decision models desiring goal is to obtain rules, which are both accurate and comprehensible. The fitness function evaluates the quality of each rule. Majority of fitness functions used in GP grew out of basic concepts on classification rule evaluation. They usually use TP, FP, TN and FN factors which, for a given rule of the form ***IF condition (A) THEN consequent (B)***, denote respectively:

- TP (True Positives) – a number of examples satisfying both A and B,
- FP (False Positives) – a number of examples not satisfying A but satisfying B,
- FN (False Negatives) – a number of examples not satisfying both A and B,
- TN (True Negativess) – a number of examples satisfying A but not B.

To rely on these factors several formulas used in fitness function have been described in the literature [14], i.e.,

- confidence: ($co = TP/(TP + FP)$
- consistency ($cs = TP/(TP + TN)$
- sensitivity ($Se = TP/(TP + FN)$)
- specificity ($Sp = TN/(TN + FP)$

A fitness function usually contains combination of such formulas, e.g., $fitness = Se * Sp$ [19]. Selection of factors appearing in the fitness function clearly depends on the type of the task and desired characteristics of the discovered solution. In the Pittsburgh approach, where one individual is the entire solution, the most popular method to evaluate solutions is accuracy rate, which is simply the ratio of the number of correctly classified test examples over the total number of test examples.

Adjusting properly the fitness function can also addressed the problem of the comprehensibility of the discovered rules. Beside the accuracy of the classification rules on the data set, the function can also take into account a size of the tree, penalising the trees with the high number of nodes. Noda et al. introduced additionally a rule "interestingness" term in the fitness function in order to discover interesting rules [15].

5.3.2 An Example of Classification Rules Discovering using GP

This section contains a description of a successful application of GP to a classification task. As an example we propose research of Ross, Fueten and Yashir [16]. They used GP to solve the problem of identification of microscopic-sized minerals involved in rocks. Mineral identification is a task, that usually done manually by an expert, who uses various visual cues like colour, texture and the interaction of the crystal lattice with different directions of polarised light. It is worth to stress, that proper identification by a human is time-consuming and requires years of experience.

GP algorithm is used by Ross et al. to develop a decision tree from database, which contain 35 colour and texture parameters of mineral specimens' feature. This database was used also to manual mineral identification. These parameters were obtained by multiple-stage process of mineral sample images analysis. Their approach is a Michigan approach, so the algorithm produces one decision tree specialised for identifying a specific mineral. Such a tree, given a sample of data for a grain, can determine, if the grain corresponds to the mineral, the tree is engineered to recognise. Thus decision model consists of several decision trees, one for each mineral species.

5.3.3 Used Algorithm

To represent solution two languages were proposed, so it is a Grammar-Based Genetic Programming algorithm. The Backus-Naur grammar for the languages is as below:

```
L1: Bool     ::=if Param < Param then Bool
                |Param<Param |True|False
    Param    ::=p1|p2|...|p35|<ephemeral const>
L2: Bool     ::=if Bool then Bool else Bool|Bool and Bool
                |Bool or Bool|not Bool|E BoolRel E|True
                |False
    BoolRel ::=<|<=|>|>=|=
    E        ::=E Op E|max(E1,E2)|min(E1,E2)|Param
    Op       :=+|-|/|*
    Param    ::=p1|p2|...|p35<ephemeral const>
```

The first language **L1** consists only of ***If-Then*** statements, a relational operator "<" over the 34 parameter values and the logical constants *True* and *False*, while **L2** is more complex and introduces mathematical expressions over parameter values. Ephemeral constants in **L2** are floating-point numbers between 0.0 and 1.0 and relational "<" statement work with a degree of error within 10% of the left-hand parameter. For example, in the expression "a=b", if b is within 10% of value of a, the expression returns *True*. Additionally, for **L2** language it was necessary to modify division operator "/" to satisfy closure property. Thus this operator returns "0", if division by zero appears. The individual coded in both languages return *True*, if the analysed grain parameters conform to the particular mineral, that the tree is designed to identify, and *False* otherwise.

The data set for evolution and experiments contained 11 mineral species and varied number of examples of each mineral: quartz (575), K-spar (369), plag (358), biotite (74), muscovite (36), hornblende (86), CPX (81), olivine (685), garnet (311), calcite (520) and opaque (189). The data set was prepared with participation of geologist. For a given mineral selected for one evolution process, training and testing data are chosen from the entire data set. The training data set consists of a random selection of K samples of mineral being identified and a random selection of K samples for each of the other minerals, so the total number of training examples is N^*K, where N is the number of mineral species.

The fitness function used during evolution for a distinguished mineral M is the following:

$$Fitness = \frac{1}{2}K\left(\frac{Hits}{K} + \frac{Miss}{(N-1)}\right) \tag{5.4}$$

where $Hits$ is the hits correct number – the number of correctly identified grains of mineral M and $Miss$ is the misses correct number – the number of grains correctly identified as not being mineral M, K – a number of samples of M in the training set.

Experiments were carried out for two kinds of training sets. The first one contained 34 positive examples and 340 negative examples per run, in the second one the number of positives examples were doubled to 70. For experiments of the second type four of the eleven minerals (biotite, muscovite, hornblende and CPX) had to be excluded, because they provided insufficient numbers of positive examples in the grain database to permit accurate training and testing. The testing set was the rest of the database not used for training. The testing formula was similar to the fitness formula:

$$Performance = \frac{1}{2}\left(\frac{Hits}{P} + \frac{Miss}{Q}\right) 100\% \tag{5.5}$$

where P is the total number of grains for mineral M in the testing set, and Q is the total number of other grains. Table 5.3 shows obtained results of GP

Table 5.3. Obtained result in Minerals classification [16]

		L1				**L2**				**NN**
K		34		70		34		70		74
Mineral		Test	Train	Test	Train	Test	Train	Test	Train	Test
Quartz	Best:	90.2	94.6	92.2	93.8	93.5	97.5	97.8	96.1	96.5
	Av.:	84.7	90.9	85.7	88.9	88.3	93.7	90.1	92.5	
K-spar	Best:	94.2	97.8	93.9	96.0	93.8	99.0	93.4	96.9	88.6
	Av.:	89.7	95.9	91.0	93.5	90.7	97.8	90.8	93.0	
Plag	Best:	86.2	91.8	84.7	93.8	83.3	95.7	95.4	94.3	89.1
	Av.:	75.8	88.0	78.9	88.9	78.6	91.8	80.2	93.0	
Biotite	Best:	98.1	98.1	–	–	96.6	99.4	–	–	97.3
	Av.:	93.7	96.4			92.7	98.0			
Muscovite	Best:	98.3	6.6	–	–	97.6	99.9	–	–	–
	Av.:	74.8	94.1			78.0	98.0			
Homblende	Best:	91.4	97.7	–	–	92.8	98.5	–	–	95.4
	Av.:	86,8	94.7			86.1	95.8			
CPX	Best:	90.3	96.5	–	–	93.8	98.5	–	–	97.5
	Av.:	81.0	88.1			84.9	92.8			
Olivine	Best:	90.0	94.4	89.9	93.8	91.4	96.6	93.2	98.1	92.8
	Av.:	83.0	88.6	85.6	90.0	84.4	92.6	89.4	95.1	
Garnet	Best:	98.2	99.7	99.6	99.6	97.8	100	99.4	99.8	97.4
	Av.:	96.0	98.3	97.6	98.1	95.8	98.9	97.1	98.6	
Calcite	Best:	92.1	97.4	91.8	95.2	92.8	99.0	95.1	98.2	96.0
	Av.:	84.4	93.6	88.5	93.0	86.2	98.3	91.2	95.9	
Opaque	Best:	97.7	99.9	97.7	99.9	98.3	100	98.1	100	95.2
	Av.:	93.6	98.9	96.6	99.4	95.2	99.3	96.5	99.7	

runs and, for comparison, the results of testing performance for the same data by neural network (NN) proposed by Thompson et al. [17], trained to identify all the minerals with one single network.

Experiments showed that GP was successfully applied to mineral classification task. The overall performance of the best obtained mineral identifiers ranged from 86% to 98%. The authors noticed that the language **L2**, which is more complex than **L1**, did not lend any great advantages to the quality of solutions but is computationally more expensive due to arithmetic operators. Solutions obtained using more positive grain examples was normally advantageous to the average quality of solutions, but in most of the cases for **L1**, the solutions for $K = 70$ were not significantly better than those for $K = 34$ and often suffered from over-training effects. The GP performance is comparable to that obtained with a performance of NN, but obtained models are not a "black-box" and are quite short and possible to understand by a human.

5.3.4 General Remarks

Many companies collect vast amounts of data, however they fail to extract necessary information to support managerial decision-making. That is why classification is an important problem extensively studied in several research areas. Classification systems can find in databases meaningful relationships useful for supporting a decision making process or automatic identifying objects in medicine, astronomy, biology etc., that might take years to find with conventional techniques.

GP was used for deriving classification models from data especially in medical and financial domains. As an example the Gray's and Maxwell work on using GP to classify tumours based on H Nuclear Magnetic Resonance spectra can be mentioned [12]. They proposed a new constrained-syntax Genetic Programming and reported, that the algorithm can classify such data well. Other example could be one of the newest researches made by Bojarczuk et. al., who proposed new constrained-syntax Genetic Programming algorithm for discovering classification rules [12]. They tested their algorithm on five medical data sets such as: Chest pain, Ljubljana breast cancer, Dermatology, Wisconsin breast cancer and Pediatric Adrenocortical Tumor and obtained good results with respect to predictive accuracy and rule comprehensibility, by comparison with C4.5 algorithm [18]. Another interesting work was presented by Bentley, who used GP to evolve fuzzy logic rules capable of detecting suspicious home insurance claims. He showed that GP is able to attain good accuracy and intelligibility levels for real, home insurance data, capable of classifying home insurance claims into "suspicious" and "non-suspicious" classes [19].

Usually databases considered in data mining (DM) tend to have gigabytes and terabytes of data. The evaluation of individual against the database is the most time consuming operation of the GP algorithm. But thanks to variety of possible modifications and parallel approaches to Genetic Algorithms, scalability of these algorithms can be achieved.

5.4 GP for Prediction Task and Time Series Odelling

The amount of data stored in different databases continues to grow fast. This large amount of stored data potentially contains valuable, hidden knowledge. This knowledge could be probably used to improve the decision-making process. For instance, data about previous sales of a company may contain interesting relationships between products and customers, what can be very useful to increase the sales. In medical area, the probability of presence of particular disease in a new patient could be predicted by learning from historical data. In the industry field, reducing fabrication flaws of certain product can be achieved by processing the large quantity of data collected during the fabrication process. That is why analysing data to predict future outcomes is one of the most studied fields in data mining.

In prediction task the database often has a form of time series, since data are collected daily, and intuitively in many areas the order of data is of great importance. Although neural networks (NN) have been very successful in developing decision models, it is difficult to understand such models, since they are normally represented as a 'black box'. It causes that a Genetic Programming becomes popular in prediction task too.

5.4.1 Adaptation of GP to Prediction Task

In prediction tasks the goal is to predict the value (the class) of a user-specified goal attribute based on the values of other attributes. Classification models described before are often used as the prediction tool. Classification rules can be considered as a particular kind of prediction rules, where the rule antecedent contains a combination – typically, a conjunction – of conditions on predicting attribute values, and the rule consequent contains a predicted value for the goal attribute. Therefore GP algorithms for prediction purpose are usually similar to those designed for classification tasks.

GP algorithms become more complicated when learning database contains time series. The simples modification is to allow in terminal set attribute values for not only one evaluated sample, but for some previous samples from a given window size. Achieving good results however often demands further improvements. To deal with specific nature of time series data special functions are added to language or grammar, according to which individuals are generated, to allow past values as input to the evolved program or equation. These can be simple arithmetic operations like an average or maximum value of attribute, which have an implicit time range, more complex like the function: *p(attr,num)*, representing the value of *attr* for the day *num* before the current day being evaluated, or some typical for a given problem including characteristic domain knowledge. GP algorithm to time series modelling can also remain the same as for classification task but only data set can be modified during preprocessing phase, for example by converting to a time series of period-to-period percent change of each attribute.

Desirable property of such a rule is its predicting power in assigning (predicting) the class of the new object. The prediction quality is verified on the test data. Sometimes test data set is given in advance, but when there is only one data set available some techniques for prediction testing are necessary. The most commonly one is *N-fold cross validation*. This method divides the data set to N mutually exclusive data sets of the same size. In each step of total N steps of the algorithm, one sub-data set is used for the testing, and the rest of them for learning process. In the learning part of the algorithm, the goal attribute values are available, while in the testing part these values are used for evaluation of the predicting ability of the rule.

5.4.2 Adaptation of GP to Prediction Tasks

Because one of the most interesting area for prediction application is the world of finance, as an example was chosen a system to stock market analysis proposed by Myszkowski [25]. His algorithm GESTIN (GEnetic STrategies of INvestment) was designed to learn a kind of decision model – an effective investment strategy (portfolio management).

The database that was used in experiments came from Warsaw Stock Exchange and consists of 27 major indexes of stock markets. The sample period for the indexes runs from April 14, 1991 until September 7, 2001. The original series are sampled at daily frequency. The sample periods correspond with 2147 observations. Database contains indexes for 223 join-stock companies.

The first modification introduced to GESTIN comparing with the standard GP algorithm is the form of an individual. An individual represents a decision model, which consists from 2 trees. One tree determines, if given stocks should be bought, and the other, if stocks should be sold. Inner nodes in trees are logical operators: AND, OR and NOT. Outer nodes (terminals) however are not a simple values but complex expressions, which contain indexes' names (names of attributes from database), arithmetical operators and constants of different type. Table 5.4 shows names of indexes and syntax of contained them outer nodes. Fig. 5.4 shows an example of a decision tree. If the value

Table 5.4. Indexes in outer nodes

Name of index (NoI)	Syntax of leaf
INX, MA8, MA9, MA12, MA15, MA17, MA26, CK	NoI Opeartor Float+ NoI Operator Float+ * NoI
MACD8, MACD12	NoI Operator Long
WOL, WRB, WRO, WRN, AT	NoI Operator Long
EPS, FE, PE, ROA, ROE, RS, ROC, OBV	NoI Operator Float
EPSR, FER, RSI, MAN	NoI Operator $< -100..100 >$

of boolean expression coded in that tree is true, it means (depending on kind of the tree) to sell or to buy stock of a given company.

The course of the algorithm (Fig. 5.5) is quite typical. Firstly a random population is generated and if termination criteria are not met, evaluation, selection, crossover and mutation operations on individuals are performed. Characteristic element of the evolution process is the evaluation phase, which

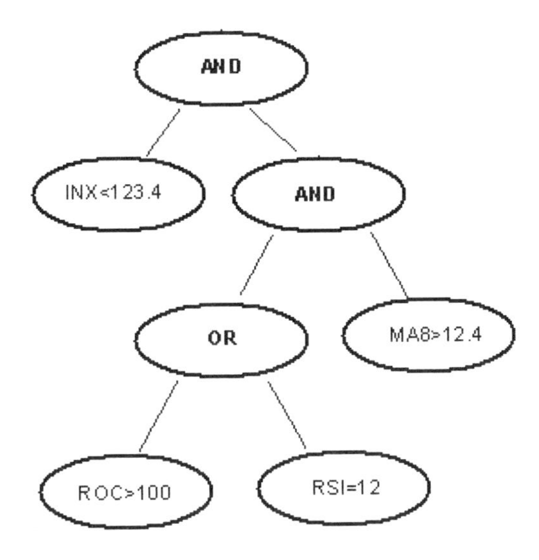

Fig. 5.4. An exemplary tree in GESTIN algorithm

1. Create an initial GP population
2. While termination criteria not met do:
3. Evaluate the fitness of the population members
4. For each session from database do (from the oldest to the newest one):
5. For each join-stock company C do:
6. For each individual I do:
7. If I possesses stock of company C and I wants to sell V:
8. sell V
9. If I doesn't possess stock of company C and I wants to by V:
10. buy V
11. Notice changes in individual's portfolio
12. Sell all stock from portfolio of individual
13. Make ranking of individuals
14. Move N best individuals to the next generation unchanged
15. Probabilistically apply crossover operator
16. Probabilistically apply mutation operator
17. Go to step 2.
18. End.

Fig. 5.5. The scheme of GESTIN algorithm

needs closer insight. An individual is allowed to speculate on the stock exchange by buying and selling stocks securities sequentially starting with an empty portfolio for data from each session in database from the oldest to the newest session. Finally, all stocks are sold and ranking of individuals is performed according to gain or loss given strategy brought. As the speculation process in reality is very complex, some restriction and simplification are applied during the evaluation process. An individual can't buy stock of company he already possesses. The size of individual's portfolio is set in advance. Additionally, each individual has the same amount of ready cash (rc) at the beginning and maximal number of possible companies (nc) is set, but an individual can spend on stock of one company only rc/nc of cash it has at the beginning, to insert portfolio's diversification. To make speculation more realistic a commission of stockbroker parameter is added, which significantly affect the value of the fitness function.

It is not difficult to create strategy, which is very good for data, it was learn on. It is difficult to create a strategy earning regardless of time period. To check prediction ability of models obtained by GESTIN, some experiments were performed. Firstly 4 different time periods were chosen: one learning – T1 and tree testing – T2, T3 and T4. These periods are characterised by different WIG index values. The first one – the learning period – is a time of bull market (WIG=+13%), the second is a time of bear market (WIG=–28%), the third one is a mixed period of bull market and bear market (WIG=0%), and the last one is a bear market with elements of bull market. Index WIG is the main index in Warsaw Stock Exchange, and it reflects results potentially obtained using the simplest strategy: "buy and keep" consisting in buying stock at the beginning and selling in the end. Results for using 10 different obtained strategies to the 4 time periods are shown in table 5.5.

Table 5.5. Results of 10 different strategies for one learning and three testing time periods of different value of WIG index

	T1	**T2**	**T3**	**T4**
WIG	+13%	–28%	+0%	–8.9%
Strategy 1	+23%	–15%	+0.8%	+6.8%
Strategy 2	+21%	–23%	–4%	+10%
Strategy 3	+21%	–23%	–4%	+4.8%
Strategy 4	+19.4%	–22%	–20.7%	+4.4%
Strategy 5	+25%	–23%	+2.8%	+7.4%
Strategy 6	+18%	–15%	–16%	+3.6%
Strategy 7	+21.6%	–3.2%	+10%	–8.4%
Strategy 8	+21.6%	–21.2%	–4%	+11.6%
Strategy 9	+19.5%	–11%	+9.1%	+2.6%
Strategy 10	+23.8%	–16%	+4.5%	+5.5%
Average	**+21%**	**–19%**	**–2 %**	**+5%**

Experimental results turned out to be quite good. Except from the time period T3 average results are better than index WIG. But the strategies are far away from allowing gain during bear market. The best one seems to be the 9-th strategy, which has quite simple and short notation:

```
Selling tree: MACD12<-3.8394
Buying tree : (WOL<-14882 OR RSI<14.43) OR FE>3.3286e+08
```

5.4.3 Hybrid GP and GA to Develope Solar Cycle's Model

The next example shows how the hybrid GP+AG algorithm for mathematical modelling (described in section 5.2.2) can be used to inferring models based on time series data. The algorithm was used to develope solar cycle's model and experiments concerned both approximating and predicting abilities of the algorithm were conducted.

Sunspots appear as dark spots on the surface of Sun. They typically last for several days, although very large ones may live for several weeks. It is known for over 150 years that sunspots appear in cycles. The average number of visible sunspots varies over time, increasing and decreasing on a regular cycle of between 9.5 to 11 years, on average about 10.8 years. The part of the cycle with low sunspot activity is referred to as "solar minimum" while the portion of the cycle with high activity is known as "solar maximum". The number of sunspots during solar maximum various for cycle to cycle. They are collected for about 200 years. Changes in the sun activity are shown in Fig. 5.6. Knowledge of solar activity levels is often required years in advance, e.g., during satellite orbits and space missions planning [20].

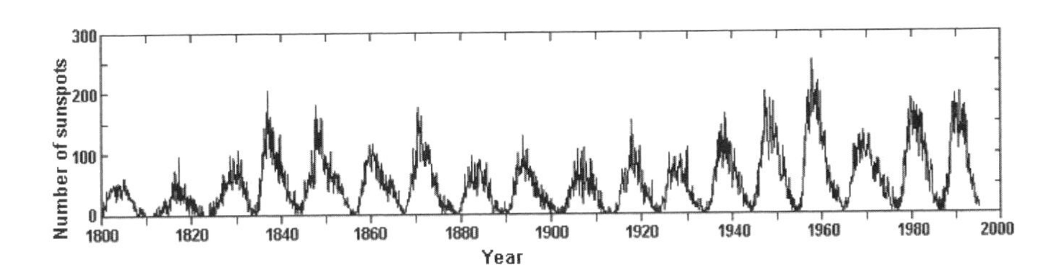

Fig. 5.6. Changes in Sun activity

To run the algorithm, originally daily collected data had to be specifically transformed into database suited to the GP. General way in which training and test data were generated is presented in Fig. 5.7. Firstly the learning time period (t_0, t_1), the number of attributes n, and time period length d_1 is determined. Attribute values are sum of sunspots appearing during time period of length d_1. Sunspot number for attribute y is counted for time period length d_2, different or equal to d_1, see Fig. 5.7a). Additionally distance p,

denoted distance between time periods used to value of nth attribute and y, is set. According to this, if $s(t_p, t_k)$ means sunspots number for the time period (t_p, t_k), where t_p is the initial and t_k the final date, for the learning time period (t_0, t_1) attribute values for the first training sample in database is as follows:
$x_1 = s(t_0, (t_0 + d_1))$,
$x_2 = s((t_0 + d_1), (t_0 + 2d_1)), ...,$
$x_n = s((t_0 + (n-2)d_1), (t_0 + (n-1)d_1))$,
$y = s((t_0 + (n-1)d_1 + p), (t_0 + (n-1)d_1 + p + d_2))$
The second training sample is generated similarly, but t_0 gets value $t_0 + d_1$. The last k^{th} sample is the one for which y is set for time period $(t_1, t_1 + d_2)$. Training database is generated in the same way but the initial test sample is the one with first attribute value counted for a time period $(t_1, t_1 + d_2)$, and d_1, d_2, n, p parameters remind unchanged.

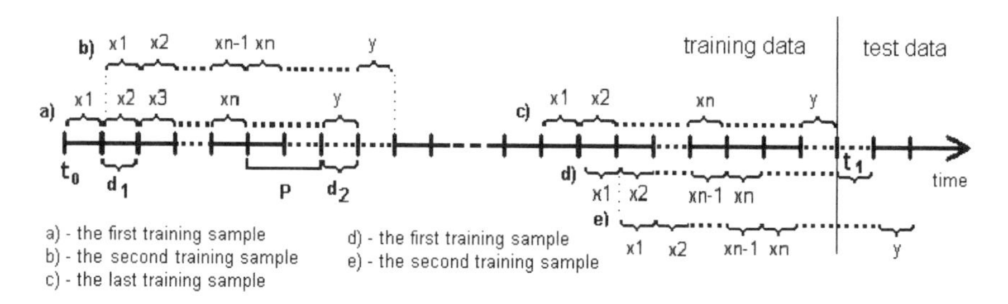

Fig. 5.7. Data preprocessing

In short term tests approximation turned out to be very good. Fig. 5.8 and 5.9 show results of two short-time experiments (for different values of parameters) as a comparison between value of an attribute y of given samples from training set and value calculated with obtained model. Evolution was led during 900 generations. In long-term tests approximation differed from one time period to another but was still surprisingly good. The best approximation for the given values of parameters (presented in Fig. 5.10) was obtained for $d_1 = d_2 = 180$ days, $p = 900$ days and learning time period from 1914 to 1980 year. Average percentage error obtained during this experiment was equal 16.9%.

Although derived functions well reflected relationships between data, they turned out to be useless at predicting future values of solar activity both in short and long advance. Increasing number of attributes to 30-40 and lengthening the learning time period could probably improve predicting abilities, but in the case of solar activity, it is not possible due to lack of data collecting in previous centuries.

Described experiments showed that GP algorithm can perform well in real-data, time series modelling. But the prediction task, often much more useful

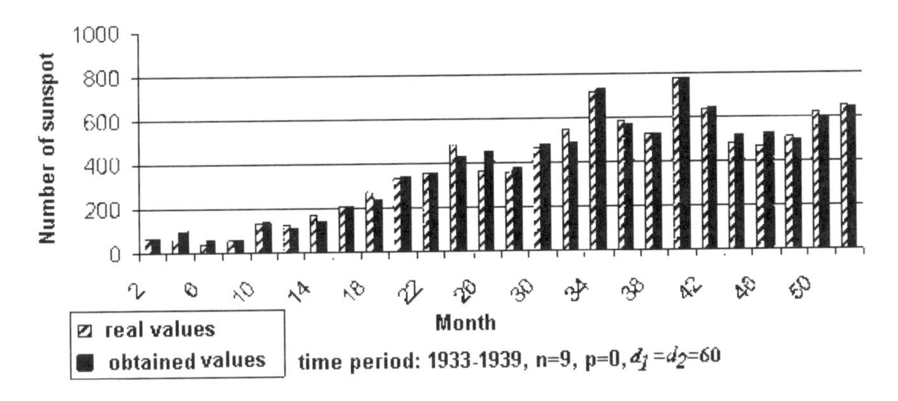

Fig. 5.8. Sunspot approximation – short-time test

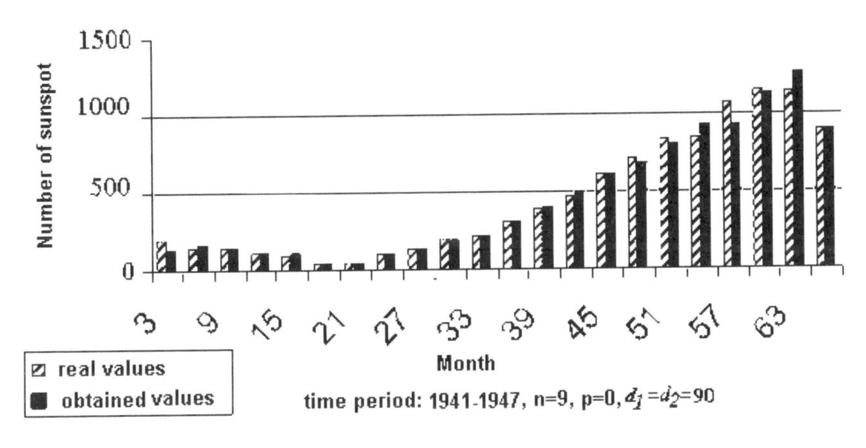

Fig. 5.9. The best approximation of sunspots – short-time test

and desired, is however much more difficult and data-dependent, so in this case the GP algorithm failed.

5.4.4 General Remarks

Advances in data collection methods, storage and processing technologies are provided a unique challenge and opportunity for automated data exploration techniques. Enormous amounts of data are being collected periodically: daily, weekly, monthly from great scientific projects like the Human Genome Project, the Hubble Space Telescope, from stock trading, from computerised sales and other sources. GP was applied to variety of time series modelling problems for understanding the dynamics of natural systems or/and for decision support systems development.

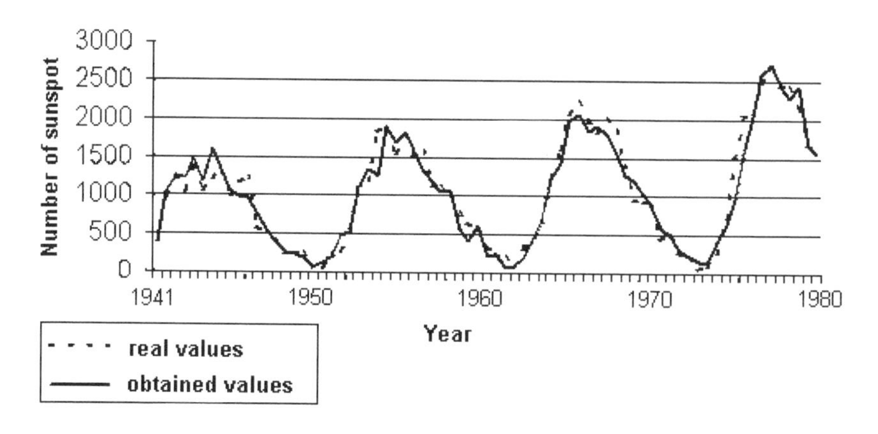

Fig. 5.10. The best approximation of sunspots – long-time test

Each approach demanded introducing of special elements to the standard GP algorithm. As an example could be mentioned Bhattacharyya, Pictet, and Zumbach work, who induce trading decision models from high-frequency foreign exchange FX markets data. They suggested incorporation of domain-related knowledge and semantic restriction to enhance GP search. Their trading model seeks to capture market movement patterns and thereby provides trading recommendations in the form of signals, a " 1" indicating a buy signal, " –1" a sell signal, and "0" indicating to stay out of the market [21]. Another example in financial domain is Doherty's system. Doherty applied GP for induction of useful classification rules from the Standard and Poors COMPUSTAT database, which contains 334 attributes, spanning 50 years of data for nearly 10,000 active (trading on public markets) and 11,000 inactive (non-trading, acquired, or failed) public companies from North America. He transformed and normalised the database for better performance of evolution process [23]. GP was also used in non-financial domains, for example to ecological time series analysis. Whigham's and Recknagel's work on applying the Grammar Based Genetic Programming framework for the discovery of predictive equations and rules for phytoplankton abundance in freshwater lakes from time series data. They have demonstrated that models can be developed for the non-linear dynamics of phytoplankton, both as a set of rules and as mathematical equations [23].

But, what can be noticed analysing existing approaches, it is easy to obtain model exactly (or almost exactly) reflecting relationships hidden in the training database, but in a case of predicting – accuracy of obtained models "good results" are consider to be those, for which prediction on the validation set is even slightly more right than wrong. This shows how difficult prediction task is, and how much further work is needed.

5.5 Summary

In the chapter we can see usefulness of Genetic Programming for solving selected problems. These problems are connected with data processing – mathematical modelling the numerical data, classification task on the basis of possessed examples, prediction and time series modelling. The problems are not new ones, but still there are not commonly accepted methods for solving them. Genetic Programming is one of valuable approaches for them. In the mathematical modelling, described in section 5.2, the hybrid method – the Genetic Programming for trees developing and the classical Genetic Algorithm for parameters tuning – work together manifesting their high usefulness. The same method used for time series modelling (section 5.4) works very well for data approximation but the prediction of Sun spots is too difficult. We do not know, if there exists a solution of this problem, it is possible that this problem cannot be solved on the basis of earlier data or proposed representation of individuals.

Of course, one can find a lot of different applications of GP to data modelling. The authors try to select the interesting ones, showing different problems with using GP. As it is known, the GP is able to solve real problems – see work of Koza, e.g., aerial designed using GP is patented. But, as it is characteristic for all metaheuristics, every using of such approaches must be very carefully tuned for the solved problem. The readers are asked do not use any 'proper' values of parameters and genetic operators, it is necessary to develop the proper ones for the considered problem.

References

1. Keijzer M (2002), Scientific Discovery Using Genetic Programming. PhD Thesis, Danish Technical University, Lyngby
2. Koza JR (1992), Genetic Programming: On the Programming of Computers by Means of Natural Selection. MIT Press
3. Koza JR (1991), A genetic approach to econometric modelling. In: Bourgine P, Walliser B (eds), Economics and Cognitive Science. Pergamon Press, Oxford
4. Montana DJ (1995), Strongly typed genetic programming. Evolutionary Computation 3(2):199-230
5. Raidl GR (1998), A Hybrid GP Approach for Numerically Robust Symbolic Regression. Proc. of the 1998 Genetic Programming Conference, Madison, Wisconsin
6. Sugimoto M, Kikuchi S, Tomita M (2003), Prediction of Biochemical Reactions Using Genetic Programming. Genome Informatics 14: 637-638
7. Langdon WB, Barrett SJ (2004), Genetic Programming in Data Mining for Drug Discovery. In: Ghosh A, Jain LC (eds), Evolutionary Computing in Data Mining. Springer
8. Makarov DE, Metiu H (2000), Using Genetic Programming To Solve the Schrodinger Equation. The Journal of Physical Chemistry A 104 8540-8545

9. Turton I, Openshaw S, Diplock G (1996), Some Geographical Applications of Genetic Programming on the Cray T3D Supercomputer. Proc. of UKPAR'96, Jessope, Shafarenko (eds)
10. Freitas AA (2002), Data Mining and Knowledge Discovery with Evolutionary Algorightms. Springer-Verlag
11. Mallinson H, Bentley P (1999), Evolving Fuzzy Rules for Pattern Classification. Computational Integration for Modelling. Control and Automation '99, Masoud Mohammadian, IOS Press
12. Bojarczuk CE, Lopes HS, Freitas AA (2001), Data mining with constrained-syntax genetic programming: applications in medical data sets. Proc. Intelligent Data Analysis in Medicine and Pharmacology, London
13. Koza JR (1991), Concept formation and decision tree induction using the genetic programming paradigm. In: Schwefel H P, Mnner R (eds), Parallel Problem Solving from Nature. Springer-Verlag, Berlin
14. Freitas AA (2002), A survey of evolutionary algorithms for data mining and knowledge discovery. Ghosh A, Tsutsui S. (eds.) Advances in Evolutionary Computation. Springer-Verlag
15. Noda E, Freitas AA, Lopes HS (1999), Discovering interesting prediction rules with a genetic algorithm. Conference Evolutionary Computation (CEC-99), Washington
16. Ross J, Fueten F, Yashkir DY (2001), Automatic mineral identification using genetic programming. Machine Vision and Applications 13: 61-69, Springer-Verlag
17. Thompson S, Fueten F, Bockus D (2001), Mineral identification using artificial neural networks and the rotating polarizer stage. Comput Geosci, Pergamon Press
18. Gray HF, Maxwell RJ, Martinez-Perez I, Arus C, Cerdan S (1996), Genetic programming for classification of brain tumours from nuclear magnetic resonance biopsy spectra. In: Koza J R, Goldberg D E, Fogel D B, Riolo R L (eds), Proc. of the First Annual Conference, Stanford University, MIT Press
19. Bentley PJ (2000), "Evolutionary, my dear Watson" – investigating committee-based evolution of fuzzy rules for the detection of suspicious insurance claims. Proc. Genetic and Evolutionary Computation Conf. (GECCO-2000), Morgan Kaufmann
20. Kippenhahn R (1994), Discovering the Secrets of the Sun. John Wiley & Sons
21. Bhattacharyya S, Pictet O V, Zumbach G (2002), Knowledge-Intensive Genetic Discovery in Foreign Exchange Markets. IEEE Transactions on Evolutionary Computation
22. Doherty CG (2003), Fundamental Analysis Using Genetic Programming for Classification Rule Induction. Genetic Algorithms and Genetic Programming at Stanford 2003, Stanford Bookstore
23. Whigham PA, Recknagel F (2001), An Inductive Approach to Ecological Time Series Modelling by Evolutionary Computation. Ecological Modelling
24. Szpunar E (2001), Data mining methods in prediction of changes in solar activity (in polish). MA Thesis, Wroclaw University of Technology
25. Myszkowski PB (2002), Data mining methods in stock market analysis (in polish). MA Thesis, Wroclaw University of Technology

6

Stock Market Modeling Using Genetic Programming Ensembles

Crina Grosan[1] and Ajith Abraham[2]

[1] Department of Computer Science, Faculty of Mathematics and Computer Science, Babeş Bolyai University, Kogalniceanu 1, Cluj-Napoca, 3400, Romania. cgrosan@cs.ubbcluj.ro, http://www.cs.ubbcluj.ro/ cgrosan

[2] School of Computer Science and Engineering, Chung-Ang University, 221, Heukseok-Dong, Dongjak-Gu, Seoul 156-756, Korea ajith.abraham@ieee.org, http://ajith.softcomputing.net

The use of intelligent systems for stock market predictions has been widely established. This chapter introduces two Genetic Programming (GP) techniques: Multi-Expression Programming (MEP) and Linear Genetic Programming (LGP) for the prediction of two stock indices. The performance is then compared with an artificial neural network trained using Levenberg-Marquardt algorithm and Takagi-Sugeno neuro-fuzzy model. We considered Nasdaq-100 index of Nasdaq Stock Market and the S&P CNX NIFTY stock index as test data. Empirical results reveal that Genetic Programming techniques are promising methods for stock prediction. Finally formulate an ensemble of these two techniques using a multiobjective evolutionary algorithm. Results obtained by ensemble are better than the results obtained by each GP technique individually.

6.1 Introduction

Prediction of stocks is generally believed to be a very difficult task. The process behaves more like a random walk process and time varying. The obvious complexity of the problem paves way for the importance of intelligent prediction paradigms. During the last decade, stocks and futures traders have come to rely upon various types of intelligent systems to make trading decisions [1], [2], [7]. This chapter presents a comparison of two genetic programming techniques (MEP and LGP), an ensemble MEP and LGP, artificial neural network and a neuro-fuzzy system for the prediction of two well-known stock indices namely Nasdaq-100 index of Nasdaq^{SM} [19] and the S&P CNX NIFTY stock index [20]. Nasdaq-100 index reflects Nasdaq's largest companies across major

C. Grosan and A. Abraham: *Stock Market Modeling Using Genetic Programming Ensembles*, Studies in Computational Intelligence (SCI) **13**, 131–146 (2006)
www.springerlink.com

industry groups, including computer hardware and software, telecommunications, retail/wholesale trade and biotechnology [21]. The Nasdaq-100 index is a modified capitalization-weighted index, which is designed to limit domination of the Index by a few large stocks while generally retaining the capitalization ranking of companies. Through an investment in Nasdaq-100 index tracking stock, investors can participate in the collective performance of many of the Nasdaq stocks that are often in the news or have become household names. Similarly, S&P CNX NIFTY is a well-diversified 50 stock index accounting for 25 sectors of the economy [13]. It is used for a variety of purposes such as benchmarking fund portfolios, index based derivatives and index funds. The CNX Indices are computed using market capitalisation weighted method, wherein the level of the Index reflects the total market value of all the stocks in the index relative to a particular base period. The method also takes into account constituent changes in the index and importantly corporate actions such as stock splits, rights, etc without affecting the index value.

MEP and LGP techniques are applied for modeling the Nasdaq-100 and NIFTY stock market indices so as to optimize the performance indices (different error measures, correlation coefficient and so on). Results obtained by MEP and LGP are compared with the results obtained using an artificial neural network trained using the Levenberg-Marquardt algorithm [5] and a Takagi-Sugeno fuzzy inference system learned using a neural network algorithm (neuro-fuzzy model) [3][15]. Neural networks are excellent forecasting tools and can learn from scratch by adjusting the interconnections between layers. Neuro-fuzzy computing is a popular framework wherein neural network training algorithms are used to fine-tune the parameters of fuzzy inference systems. We will build an ensemble between MEP and LGP using a multiobjective evolutionary algorithm (Non-dominated Sorting Genetic Algorithm II (NSGAII)). Results obtained by ensemble are then compared with the results obtained by MEP and LGP individually.

In Section 6.1, we briefly describe the stock marketing modeling problem. In Section 6.2, different connectionist paradigms used in experiments are presented. In Section 6.3, we formulate the ensemble between MEP and LGP followed by experimentation setup in Section 6.4 and results in Section 6.5. Some conclusions are also provided towards the end.

6.2 Modeling Stock Market Prediction

We analysed the Nasdaq-100 index value from 11 January 1995 to 11 January 2002 [19] and the NIFTY index from 01 January 1998 to 03 December 2001 [20]. For both indices, we divided the entire data into almost two equal parts. No special rules were used to select the training set other than ensuring a reasonable representation of the parameter space of the problem domain. The complexity of the training and test data sets for both indices are depicted in Figures 6.1 and 6.2 respectively.

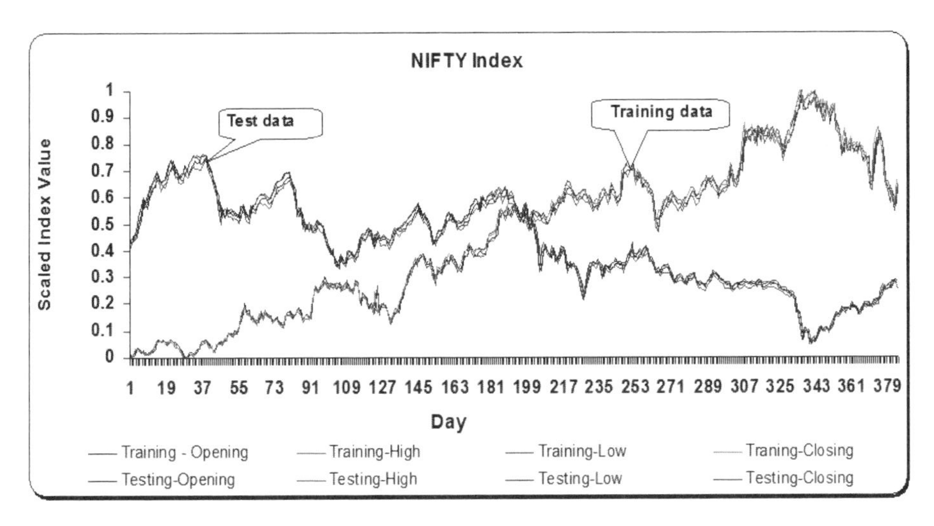

Fig. 6.1. Training and test data sets for Nasdaq-100 Index

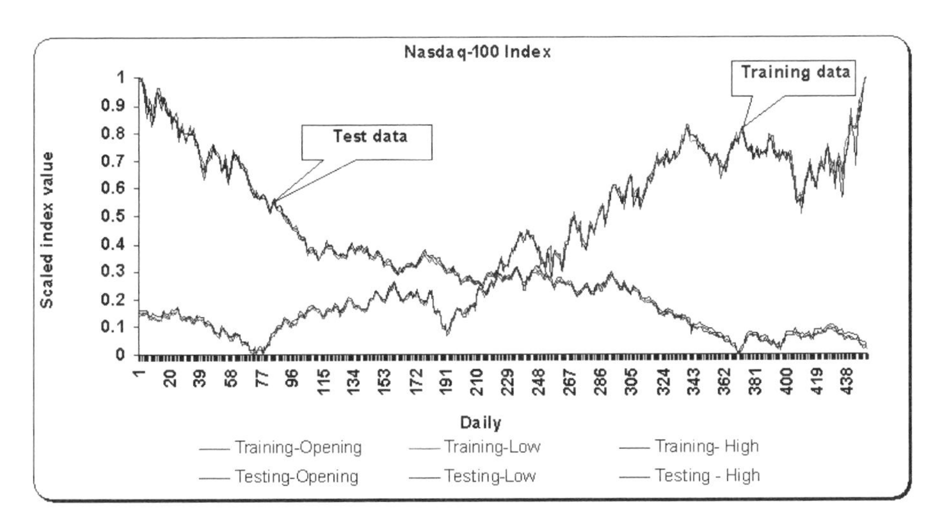

Fig. 6.2. Training and test data sets for NIFTY index

Our goal is to optimize several error measures: Root Mean Squared Error (RMSE), Correlation Coefficient (CC), Maximum Absolute Percentage Error (MAP) and Mean Absolute Percentage Error (MAPE):

$$RMSE = \sqrt{\sum_{i=1}^{N} |P_{actual,i} - P_{predicted,i}|} \quad (6.1)$$

$$CC = \frac{\sum_{i=1}^{N} P_{predicted,i}}{\sum_{i=1}^{N} P_{actual,i}}, \quad (6.2)$$

$$MAP = \max\left(\frac{|P_{actual,\ i} - P_{predicted,\ i}|}{P_{predicted,\ i}} \times 100\right) \quad (6.3)$$

$$MAPE = \frac{1}{N}\sum_{i=1}^{N}\left[\frac{|P_{actual,\ i} - P_{predicted,\ i}|}{P_{actual,\ i}}\right] \times 100, \quad (6.4)$$

where $P_{actual,i}$ is the actual index value on day i, $P_{predicted,i}$ is the forecast value of the index on that day and N = total number of days. The task is to have minimal values of RMSE, MAP and MAPE and a maximum value for CC.

6.3 Intelligent Paradigms

The different paradigms used in this chapter are described in this Section.

6.3.1 Multi Expression Programming (MEP)

MEP is a Genetic Programming variant that uses a linear representation of chromosomes. MEP individuals are strings of genes encoding complex computer programs. When MEP individuals encode expressions, their representation is similar to the way in which compilers translate *C* or *Pascal* expressions into machine code [4].

A unique MEP feature is the ability of storing multiple solutions of a problem in a single chromosome. Usually, the best solution is chosen for fitness assignment. When solving symbolic regression or classification problems (or any other problems for which the training set is known before the problem is solved) MEP has the same complexity as other techniques storing a single solution in a chromosome (such as Genetic Programming [17], Cartesian Genetic Programming [18], Gene Expression Programming [11] or Grammatical Evolution [24]).

Evaluation of the expressions encoded into a MEP individual can be performed by a single parsing of the chromosome.

Offspring obtained by crossover and mutation are always syntactically correct MEP individuals (computer programs). Thus, no extra processing for repairing newly obtained individuals is needed. (For technical details, the reader is advised to refer to the chapter on Evolving Intrusion Detection Systems or see [22].)

6.3.2 Linear Genetic Programming (LGP)

Linear genetic programming is a variant of the GP and is throughly described in chapter on Evolving Intrusion Detection Systems. LGP uses a specific linear representation of computer programs. Instead of the tree-based GP expressions of a functional programming language (like *LISP*) programs of an imperative language (like *C*) are evolved.

An LGP individual is represented by a variable-length sequence of simple *C* language instructions. Instructions operate on one or two indexed variables (registers) r, or on constants $c$from predefined sets. The result is assigned to a destination register, for example, $ri = rj* \ c$. Here is an example LGP program:

```
void LGP(double v[8]) {
    v[0] := v[5] + 73;
    v[7] := v[3] – 59;
    if (v[1] > 0)
    if (v[5] > 21)
    v[4] := v[2] .v[1];
    v[2] := v[5] + v[4];
    v[6] := v[7] .25;
    v[6] := v[4] – 4;
    v[1] := sin(v[6]);
    if (v[0] > v[1])
    v[3] := v[5] .v[5];
    v[7] := v[6] .2;
    v[5] := v[7] + 115;
    if (v[1] ≤ v[6])
    v[1] := sin(v[7]);
}
```

A LGP can be turned into a functional representation by successive replacements of variables starting with the last effective instruction. The maximum number of symbols in a LGP chromosome is 4 * Number of instructions. Evolving programs in a low-level language allows us to run those programs directly on the computer processor, thus avoiding the need of an interpreter. In this way the computer program can be evolved very quickly.

An important LGP parameter is the number of registers used by a chromosome. The number of registers is usually equal to the number of attributes of the problem. If the problem has only one attribute, it is impossible to obtain a complex expression such as the quartic polynomial. In that case we have to use several supplementary registers. The number of supplementary registers depends on the complexity of the expression being discovered.

An inappropriate choice can have disastrous effects on the program being evolved.

LGP uses a modified steady-state algorithm. The initial population is randomly generated. The following steps are repeated until a termination criterion is reached: Four individuals are randomly selected from the current population. The best two of them are considered the winners of the tournament and will act as parents. The parents are recombined and the offspring are mutated and then replace the losers of the tournament.

We used a LGP technique that manipulates and evolves a program at the machine code level. The settings of various linear genetic programming system parameters are of utmost importance for successful performance of the system. The population space has been subdivided into multiple subpopulation or demes. Migration of individuals among the sub-populations causes evolution of the entire population. It helps to maintain diversity in the population, as migration is restricted among the demes. Moreover, the tendency towards a bad local minimum in one deme can be countered by other demes with better search directions. The various LGP search parameters are the mutation frequency, crossover frequency and the reproduction frequency: The crossover operator acts by exchanging sequences of instructions between two tournament winners. Steady state genetic programming approach was used to manage the memory more effectively.

6.3.3 Artificial Neural Network (ANN)

The artificial neural network methodology enables us to design useful nonlinear systems accepting large numbers of inputs, with the design based solely on instances of input-output relationships. For a training set T consisting of $n$argument value pairs and given a d-dimensional argument x and an associated target value t will be approximated by the neural network output. The function approximation could be represented as

$$T = \{(x_i, t_i) : i = 1 : n\} \tag{6.5}$$

In most applications the training set T is considered to be noisy and our goal is not to reproduce it exactly but rather to construct a network function that generalizes well to new function values. We will try to address the problem of selecting the weights to learn the training set. The notion of closeness on the training set T is typically formalized through an error function of the form

$$\psi_T = \sum_{i=1}^{n} \parallel y_i - t_i \parallel^2 \tag{6.6}$$

where y_i is the network output.

Levenberg-Marquardt Algorithm

The Levenberg-Marquardt (LM) algorithm [5] exploits the fact that the error function is a sum of squares as given in (6.6). Introduce the following notation for the error vector and its Jacobian with respect to the network parameters w

$$J = J_{ij} = \frac{\partial e_j}{\partial w_i}, i = 1 : p, j = 1 : n \tag{6.7}$$

The Jacobian matrix is a large p × n matrix, all of whose elements are calculated directly by backpropagation technique. The p dimensional gradient g for the quadratic error function can be expressed as:

$$g(w) = \sum_{i=1}^{n} e_i \nabla e_i(w) = Je \tag{6.8}$$

and the Hessian matrix by:

$$\begin{aligned} H = H_{ij} &= \frac{\partial^2 \psi_T}{\partial w_i \partial w_j} \\ &= \frac{1}{2} \sum_{k=1}^{n} \frac{\partial^2 e_k^2}{\partial w_i \partial w_j} = \sum_{k=1}^{n} \left(e_k \frac{\partial^2 e_k}{\partial w_i \partial w_j + \frac{\partial^2 e_k}{\partial w_i \partial w_j}} \right) \\ &= \sum_{k=1}^{n} \left(e_k \frac{\partial^2 e_k}{\partial w_i \partial w_j} + J_{ik} J_{jk} \right) \end{aligned} \tag{6.9}$$

Hence defining $D = \sum_{i=1}^{n} e_i \nabla^2 e_i$yields the expression

$$H(w) = JJ^T + D \tag{6.10}$$

The key to the LM algorithm is to approximate this expression for the Hessian by replacing the matrix D involving second derivatives by the much simpler positively scaled unit matrix $\in I$. The LM is a descent algorithm using this approximation in the form below:

$$M_k = \left[JJ^T + \in I \right]^{-1}, w_{k+1} = w_k - \alpha_k M_k g(w_k) \tag{6.11}$$

Successful use of LM requires approximate line search to determine the rate α_k. The matrix JJ^T is automatically symmetric and non-negative definite. The typically large size of J may necessitate careful memory management in evaluating the product JJ^T. Hence any positive $\in$ will ensure that M_k is positive definite, as required by the descent condition. The performance of the algorithm thus depends on the choice of $\in$.

When the scalar $\in$ is zero, this is just Newton's method, using the approximate Hessian matrix. When $\in$ is large, this becomes gradient descent with a small step size. As Newton's method is more accurate, $\in$ is decreased after each successful step (reduction in performance function) and is increased only when a tentative step would increase the performance function. By doing this, the performance function will always be reduced at each iteration of the algorithm.

6.3.4 Neuro-Fuzzy System

Neuro Fuzzy (NF) computing is a popular framework for solving complex problems [3]. If we have knowledge expressed in linguistic rules, we can build a Fuzzy Inference System (FIS) [8], and if we have data, or can learn from a simulation (training) then we can use ANNs. For building a FIS, we have to specify the fuzzy sets, fuzzy operators and the knowledge base. Similarly for constructing an ANN for an application the user needs to specify the architecture and learning algorithm. An analysis reveals that the drawbacks pertaining to these approaches seem complementary and therefore it is natural to consider building an integrated system combining the concepts. While the learning capability is an advantage from the viewpoint of FIS, the formation of linguistic rule base will be advantage from the viewpoint of ANN. Figure 6.3 depicts the 6-layered architecture of multiple output neuro-fuzzy system implementing a Takagi-Sugeno fuzzy inference system. For technical details, the reader is advised to consult [15].

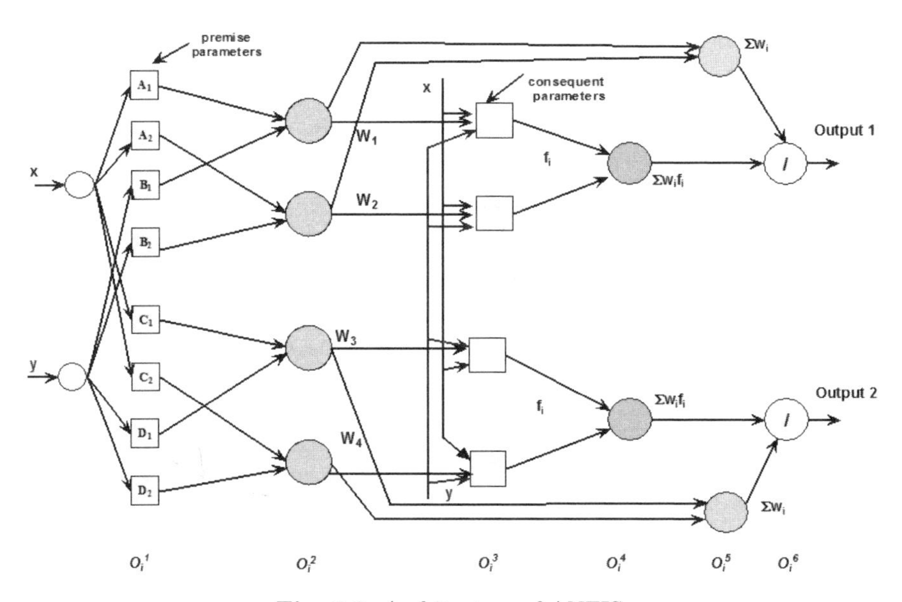

Fig. 6.3. Architecture of ANFIS

6.4 Ensemble of GP Techniques

Our goal is to optimize four error measures namely Root Mean Squared Error (RMSE), Correlation Coefficient (CC), Maximum Absolute Percentage

Error(MAP) and Mean Absolute Percentage Error (MAPE). The task is to have minimal values of RMSE, MAP, MAPE and a maximum value for CC. The objective is to carefully construct the different GP models to achieve the best generalization performance. Test data is then passed through these individual models and the corresponding outputs are recorded. Suppose results obtained by LGP and MEP are $a_n$and b_n respectively and the corresponding desired value is x_n. The task is to combine a_n andb_n so as to get the best output value that maximizes the CC and minimizes the RMSE, MAP and MAPE values.

We consider this problem as a multiobjective optimization problem in which we want to find solution of this form: $(coef_1, coef_2)$ where $coef_1$, and $coef_2$ are real numbers between -1 and 1, so as the resulting combination:

$$coef_1 * a_n + coef_2 * b_n \tag{6.12}$$

would be close to the desired value x_n. This means, in fact, to find a solution so as to simultaneously optimize RMSE, MAP, MAPE and CC. This problem is equivalent to finding the Pareto solutions of a multiobjective optimization problem. For this problem, the objectives are RMSE, MAP, MAPE and CC. We use the well known Multiobjective Evolutionary Algorithm (MOEA) - Nondominated Sorting Genetic Algorithm II (NSGAII) [10] and a short description of this algorithm is given in the subsequent Section.

6.4.1 Nondominated Sorting Genetic Algorithm II (NSGA II)

In the Nondominated Sorting Genetic Algorithm (NSGA II)[10] for each solution x the number of solutions that dominate solution x is calculated. The set of solutions dominated by x is also calculated. The first front (the current front) of the solutions that are nondominated is obtained.

Let us denote by S_i the set of solutions that are dominated by the solution x^i. For each solution x^i from the current front consider each solution x^q from the set S_i.

The number of solutions that dominates $x^q$is reduced by one. The solutions which remain non-dominated after this reduction will form a separate list. This process continues using the newly identified front as the current front. Let $P(0)$ be the initial population of size N. An offspring population $Q(t)$ of size N is created from current population $P(t)$. Consider the combined population R(t) = P(t) ∪Q(t).

Population $R(t)$ is ranked according to nondomination. The fronts F_1, $F_2, \ldots$ are obtained. New population $P(t+1)$ is formed by considering individuals from the fronts F_1, F_2, ..., until the population size exceeds N. Solutions of the last allowed front are ranked according to a crowded comparison relation.

NSGA II uses a parameter (called *crowding distance*) for density estimation for each individual. Crowding distance of a solution x is the average side-length of the cube enclosing the point without including any other point in the population. Solutions of the last accepted front are ranked according to the crowded comparison distance

NSGA II works follows. Initially a random population, which is sorted based on the nondomination, is created. Each solution is assigned a fitness equal to its nondomination level (1 is the best level). Binary tournament selection, recombination and mutation are used to create an offspring population. A combined population is formed from the parent and offspring population. The population is sorted according to the nondomination relation. The new parent population is formed by adding the solutions from the first front and the followings until exceed the population size. Crowding comparison procedure is used during the population reduction phase and in the tournament selection for deciding the winner.

6.5 Experiment Results

We considered 7 year's month's stock data for Nasdaq-100 Index and 4 year's for NIFTY index. Our target is to develop efficient forecast models that could predict the index value of the following trading day based on the opening, closing and maximum values of the same on a given day. For the Nasdaq-100index the data sets were represented by the 'opening value', 'low value' and 'high value'. NIFTY index data sets were represented by 'opening value', 'low value', 'high value' and 'closing value'. The assessment of the prediction performance of the different connectionist paradigms and the ensemble method were done by quantifying the prediction obtained on an independent data set.

6.5.1 Parameter Settings

MEP Parameter Settings. Parameters used by MEP in these experiments are presented in Table 6.1. Next two experiments analyze the results obtained by MEP by considering different population sizes and different values for chromosome length. The values for the other parameters are adapted from Table 6.1. Population size was considered 150 for both test data. Average results obtained from 10 different runs and the best results obtained are presented in Table 6.2. Results obtained for different population sizes and various chromosome lengths are presented in Tables 6.3 and 6.4 respectively for both Nasdaq and Nifty test data.

As evident from Table 6.2 the best results for Nasdaq is obtained using a population of 100 individuals and for Nifty using a population of 150 individuals. Table 6.3 illustrates that the best result for both Nasdaq and Nifty is obtained using a chromosome length of 40 and 50.

Table 6.1. Values of parameters used by MEP

Parameter		**Value**
Population size	Nasdaq	100
	Nifty	50
Number of iteration	Nasdaq	60
	Nifty	100
Chromosome length	Nasdaq	30
	Nifty	40
Crossover Probability		0.9
Functions set		+, - , *, /, sin, cos, sqrt, ln, lg, $\log_2$, min, max, abs

Table 6.2. Performance measures obtained by MEP or population sizes

			Population size			
			50	**100**	**150**	**200**
Nasdaq	**RMSE**	Best	0.022	0.032	0.035	0.0168
		Average	0.024	0.027	0.03	0.022
	CC	Best	0.999	0.097	0.97	0.992
		Average	0.995	0.984	0.98	0.997
	MAP	Best	97.43	103.31	103.31	103.31
		Average	109.59	100.37	109.33	100.7
	MAPE	Best	18.13	9.02	9.08	8.69
		Average	23.32	13.58	18.8	18.23
Nifty	**RMSE**	Best	0.0187	0.163	0.015	0.0138
		Average	0.02	0.0196	0.0197	0.019
	CC	Best	0.999	0.997	0.999	0.999
		Average	0.991	0.979	0.978	0.988
	MAP	Best	38.99	31.7	27.64	30.03
		Average	53.02	42.225	41.85	48.81
	MAPE	Best	4.131	3.72	3.17	3.027
		Average	4.9	4.66	4.81	4.34

LGP Parameter Settings. Parameters values used by LGP for Nasdaq and Nifty test data are presented in Table 6.4.

Ensemble Using NSGAII Parameter Setting. Parameters values used by NSGAII for combining (ensembling) MEP and LGP are given in Table 6.5. These parameters were used for both Nasdaq and Nifty test data.

ANN and NF parameter settings. We used a feed forward neural network with 4 input nodes and a single hidden layer consisting of 26 neurons. Tanh-sigmoidal activation function was for the hidden neurons. The training using LM algorithm was terminated after 50 epochs and it took about 4 sec-

Table 6.3. Performance measures obtained by MEP for different chromosome lengths

			Chromosome length			
			20	**30**	**40**	**50**
Nasdaq	**RMSE**	Best	0.021	0.032	0.028	0.0165
		Average	0.021	0.027	0.024	0.022
	CC	Best	0.998	0.976	0.977	0.993
		Average	0.998	0.987	0.985	0.994
	MAP	Best	97.43	103.31	103.31	103.31
		Average	97.43	100.38	118.5	115.55
	MAPE	Best	18.11	9.02	8.91	8.53
		Average	18.12	13.52	18.74	15.86
Nifty	**RMSE**	Best	0.0187	0.0169	0.015	0.014
		Average	0.0193	0.023	0.0197	0.02
	CC	Best	0.999	0.990	0.999	0.992
		Average	0.994	0.977	0.98	0.981
	MAP	Best	38.99	42.98	27.64	34.87
		Average	43.37	52.1	38.78	40.67
	MAPE	Best	4.125	4.08	3.17	3.30
		Average	4.33	5.68	4.81	4.75

Table 6.4. LGP parameter settings

Parameter	**Value**
Population size	100
Mutation frequency	95%
Crossover frequency	50%
Number of demes	10
Number of constants	60

Table 6.5. Parameters values used by NSGAII for ensembling MEP and LGP

Parameter	**Value**
Population size	250
Mutation probability	0.3
Crossover probability	0.5
Number of generations	1,000
Chromosome length	30

onds to train each dataset. For the neuro-fuzzy system, we used 3 triangular membership functions for each of the input variable and the 27 *if-then* fuzzy rules were learned for the Nasdaq-100 index and 81 *if-then* fuzzy rules for the NIFTY index. Training was terminated after 12 epochs and it took about 3 seconds to train each dataset [1].

6.5.2 Comparisons of Results Obtained by Intelligent Paradigms

Table 6.6 summarizes the results achieved for the two stock indices using the five intelligent paradigms (ANN, NF, MEP, LGP and the ensemble between LGP and MEP).

Table 6.6. Results obtained by intelligent paradigms for Nasdaq and Nifty test data

	ANN	NF	MEP	LGP	Ensemble
Test results for Nasdaq					
RMSE	0.0284	0.0183	0.021	0.021	0.0203
CC	0.9955	0.9976	0.999	0.9940	1.000
MAP	481.71	520.84	97.39	97.94	96.92
MAPE	9.032	7.615	14.33	19.65	19.25
Test results for Nifty					
RMSE	0.0122	0.0127	0.0163	0.0124	0.0127
CC	0.9968	0.9967	0.997	0.995	1.000
MAP	73.94	40.37	31.7	40.02	31.9
MAPE	3.353	3.320	3.72	2.83	2.80

The combination obtained by ensemble between LGP and MEP are reported below:

Nasdaq: $0.669668 * a_n + 0.334354 * b_n$
Nifty: $0.632351 * a_n + 0.365970 * b_n$

where a_n and b_n correspond to LGP and MEP indices respectively.

As depicted in Table 6.6, for Nasdaq test data, MEP gives the best results for MAP (97.39). Also LGP gives results very close to MEP while the other techniques did not perform that well. While MEP obtained the best result for CC, the performance obtained were close to the results obtained for RMSE by the other paradigms. Results obtained by ensemble clearly outperforms almost all considered techniques. For both Nasdaq and Nifty test data, the values obtained by the ensemble for CC is 1.000.

Since the multiobjective evolutionary algorithm was used to simultaneously optimize all the four performance measures (RMSE, CC, MAP and MAPE), more than one solution could be obtained in the final population (which, in fact, means more than one combination between LGP and MEP). Some examples of the solutions obtained for Nifty is given below:

- Best value for RMSE obtained by the ensemble is 0.012375, while CC = 0.999, MAP = 36.86 and MAPE = 2.71.
- Best value for MAP obtained by the ensemble is 25.04, while RMSE = 0.0138, CC = 0.998 and MAPE = 3.05.
- Best result obtained by the ensemble for MAPE is 2.659513, while RMSE = 0.0124, CC = 0.998 and MAP = 41.44.

6.6 Summary

In this chapter, we presented five techniques for modeling stock indices. The performance of GP techniques (empirical results) when compared to ANN and NF clearly indicate that GP could play a prominent role for stock market modeling problems. The fluctuations in the share market are chaotic in the sense that they heavily depend on the values of their immediate fore running fluctuations. Our main task was to find the best values for the several performance measures namely RMSE, CC, MAP and MAPE. We applied a multiobjective optimization algorithm in order to ensemble the GP techniques. Experiment results reveal that the ensemble performs better than the GP techniques considered separately.

According to the No Free Lunch Theorem (NFL), for any algorithm, any elevated performance over one class of problems is exactly paid for in performance over another class [26]. Taking into account of the NFL theorem, it would be a rather difficult task to predict which paradigm would perform the best for different stock indices [13].

Acknowledgements

This research was supported by the International Joint Research Grant of the IITA (Institute of Information Technology Assessment) foreign professor invitation program of the MIC (Ministry of Information and Communication), South Korea.

References

1. Abraham, A. and AuYeung, A., Integrating Ensemble of Intelligent Systems for Modeling Stock Indices, In Proceedings of 7th International Work Conference on Artificial and Natural Neural Networks, Lecture Notes in Computer Science-Volume 2687, Jose Mira and Jose R. Alverez (Eds.), Springer Verlag, Germany, pp. 774-781, 2003.

2. Abraham, A. Philip, N.S. and Saratchandran, P., Modeling Chaotic Behavior of Stock Indices Using Intelligent Paradigms. International Journal of Neural, Parallel & Scientific Computations, USA, Volume 11, Issue (1&2) pp.143-160, 2003.
3. Abraham, A., Neuro-Fuzzy Systems: State-of-the-Art Modeling Techniques, Connectionist Models of Neurons, Learning Processes, and Artificial Intelligence, Springer-Verlag Germany, Jose Mira and Alberto Prieto (Eds.), Granada, Spain, pp. 269-276, 2001.
4. Aho, A., Sethi R., Ullman J., Compilers: Principles, Techniques, and Tools, Addison Wesley, 1986.
5. Bishop, C. M., Neural Networks for Pattern Recognition, Oxford: Clarendon Press, 1995.
6. M. Brameier and W. Banzhaf, A Comparison of Linear Genetic Programming and Neural Networks in Medical Data Mining, IEEE Transactions on Evolutionary Computation, 5, pp.17-26, 2001.
7. M. Brameier and W. Banzhaf, Explicit Control of Diversity and Effective Variation Distance in Linear Genetic Programming,in Proceedings of the Fourth European Conference on Genetic Programming, edited by E. Lutton, J. Foster, J. Miller, C. Ryan, and A. Tettamanzi (Springer-Verlag, Berlin, 2002).
8. Cherkassky V., Fuzzy Inference Systems: A Critical Review, Computational Intelligence: Soft Computing and Fuzzy-Neuro Integration with Applications, Kayak O, Zadeh L A et al (Eds.), Springer, pp.177-197, 1998.
9. Collobert R. and Bengio S., SVMTorch: Support Vector Machines for Large-Scale Regression Problems, Journal of Machine Learning Research, Volume 1, pages 143-160, 2001.
10. Deb, K., Agrawal, S., Pratab, A., Meyarivan, T., A fast elitist non-dominated sorting genetic algorithms for multiobjective optimization: NSGA II. KanGAL report 200001, Indian Institute of Technology, Kanpur, India, 2000.
11. C. Ferreira, Gene Expression Programming: A New Adaptive Algorithm for Solving Problems, Complex Systems, 13 (2001) 87-129.
12. E.H. Francis et al. Modified Support Vector Machines in Financial Time Series Forecasting, *Neurocomputing* 48(1-4): pp. 847-861, 2002.
13. C. Grosan and A. Abraham, Solving No Free Lunch Issues from a Practical Perspective, In Proceedings of Ninth International Conference on Cognitive and Neural Systems, ICCNS'05, Boston University Press, USA, 2005 .
14. S. Hashem. Optimal Linear Combination of Neural Networks. *Neural Network*, Volume 10, No. 3. pp. 792-994, 1995.
15. J.S.R. Jang, C.T. Sun and E. Mizutani. Neuro-Fuzzy and Soft Computing: A Computational Approach to Learning and Machine Intelligence, Prentice Hall Inc, USA, 1997.
16. T. Joachims . Making large-Scale SVM Learning Practical. *Advances in Kernel Methods - Support Vector Learning*, B. Schölkopf and C. Burges and A. Smola (Eds.), MIT-Press, 1999.
17. J. R. Koza, Genetic Programming: On the Programming of Computers by Means of Natural Selection (MIT Press, Cambridge, MA, 1992).
18. J. Miller and P. Thomson. Cartesian Genetic Programming, in Proceedings of the Third European Conference on Genetic Programming, edited by Riccardo Poli, Wolfgang Banzhaf, Bill Langdon, Julian Miller, Peter Nordin, and Terence C. Fogarty (Springer-Verlag, Berlin, 2002).

19. Nasdaq Stock Market^{SM}: http://www.nasdaq.com.
20. National Stock Exchange of India Limited: http://www.nse-india.com.
21. M. Oltean and C. Grosan. A Comparison of Several Linear GP Techniques. *Complex Systems*, Vol. 14, Nr. 4, pp. 285-313, 2004
22. M. Oltean M. C. Grosan. Evolving Evolutionary Algorithms using Multi Expression Programming. *Proceedings of The 7th European Conference on Artificial Life*, Dortmund, Germany, pp. 651-658, 2003.
23. N.S. Philip and K.B. Joseph. Boosting the Differences: A Fast Bayesian classifier neural network, *Intelligent Data Analysis*, Vol. 4, pp. 463-473, IOS Press, 2000.
24. C. Ryan C. J.J. Collins and M. O'Neill. Gramatical Evolution: Evolving programs for an arbitrary language. In *Proceedings of the first European Workshop on Genetic Programming,* Springer-Verlag, Berlin, 1998.
25. R.E. Steuer. Multiple Criteria Optimization: Theory, Computation and Applications. New York: Wiley, 1986
26. D.H. Wolpert and W.G. Macready. No free lunch theorem for search. Technical Report SFI-TR-95-02-010. Santa Fe Institute, USA, 1995.

7

Evolutionary Digital Circuit Design Using Genetic Programming

Nadia Nedjah[1] and Luiza de Macedo Mourelle[2]

[1] Department of Electronics Engineering and Telecommunications,
Engineering Faculty,
State University of Rio de Janeiro,
Rua São Francisco Xavier, 524, Sala 5022-D,
Maracanã, Rio de Janeiro, Brazil
nadia@eng.uerj.br, http://www.eng.uerj.br/~nadia

[2] Department of System Engineering and Computation,
Engineering Faculty,
State University of Rio de Janeiro,
Rua São Francisco Xavier, 524, Sala 5022-D,
Maracanã, Rio de Janeiro, Brazil
ldmm@eng.uerj.br, http://www.eng.uerj.br/~ldmm

In this chapter, we study two different circuit encodings used for digital circuit evolution. The first approach is based on genetic programming, wherein digital circuits consist of their data flow based specifications. In this approach, individuals are internally represented by the abstract trees/DAG of the corresponding circuit specifications. In the second approach, digital circuits are thought of as a map of rooted gates. So individuals are represented by two-dimensional arrays of cells. Each of these cells consists of the logic gate name together with the corresponding input signal names as the name of the output signal is fixed. Furthermore, we compare the impact of both individual representations on the evolution process of digital circuits. Evolved circuits should minimise space and time requirements. We show that for the same input/output behaviour, employing either of these approaches yields circuits of almost the same characteristics in terms of space and response time. However, the evolutionary process is much shorter with the second encoding.

7.1 Introduction

evolutionary hardware [7] is a hardware that is yield using simulated evolution as an alternative to conventional-based electronic circuit design. Genetic

N. Nedjah and L. de M. Mourelle: *Evolutionary Digital Circuit Design Using Genetic Programming*, Studies in Computational Intelligence (SCI) **13**, 147–171 (2006)
www.springerlink.com

evolution is a process that evolves a set of individuals, which constitutes the population, producing a new population. Here, individuals are hardware designs. The more the design obeys the constraints, the more it is used in the reproduction process. The design constraints could be expressed in terms of hardware area and/or response time requirements. The freshly produced population is yield using some genetic operators such as crossover and mutation that attempt to simulate the natural breeding process in the hope of generating new design that are fitter i.e. respect more the design constraints. Genetic evolution is usually implemented using genetic algorithms.

The problem of interest consists of choosing the best encoding for evolving *rapidly* efficient and creative circuits that implement a given input/output behaviour without much designing effort. The obtained circuits are expected to be minimal both in terms of space and time requirements: The circuits must be compact i.e. use a reduced number of gates and efficient, i.e. produce the output in a short response time. The response time of a circuit depends on the number and the complexity of the gates forming the longest path in it. The complexity of a gate depends solely on the number of its inputs.

The remainder of this chapter is divided in five sections. In Section 7.2, we describe the principles of evolutionary hardware. In Section 7.3, we describe the first methodology and propose the necessary genetic operators. In Section 7.4, we describe the second approach as well as the corresponding genetic operators. In Section 7.5, we compare the evolved hardware using either methodologies. Finally, we draw some conclusions.

7.2 Principles of Evolutionary Hardware Design

Evolutionary hardware [7] consists simply of hardware designs evolved using genetic algorithms, wherein chromosomes represent circuit designs. In general, evolutionary hardware design offers a mechanism to get a computer to provide a design of circuit without being told exactly how to do it. In short, it allows one to automatically create circuits. It does so based on a high level statement of the constraints the yielded circuit must obey to. The input/output behaviour of the expected circuit is generally considered as an omnipresent constraint. Furthermore, the generated circuit should have a minimal size.

Starting form random set of circuit design, which is generally called *initial population*, evolutionary hardware design breeds a population of designs through a series of steps, called *generations*, using the Darwinian principle of natural selection. Circuits are selected based on how much they adhere to the specified constraints. *Fitter* individuals are selected, *recombined* to produce offspring which in turn should suffer some *mutations*. Such offspring are then used to populate of the next generation. This process is iterated until a circuit design that obeys to all prescribed constraints is encountered within the current population.

Each circuit within the population is assigned a value, generally called *fitness.* A circuit design is it fit if and only if it satisfies the imposed input/output behaviour. A circuit design is considered *fitter* than another if and only if it has a smaller size.

An important aspect of evolutionary hardware design is thus to provide a way to evaluate the adherence of evolved circuits to the imposed constraints. These constraints are of two kinds. First of all, the evolved circuit design must obey the input/output behaviour, which is given in a tabular form of expected results given the inputs. This is the truth table of the expected circuit. Second, the circuit must have a reduced size. This constraint allows us to yield compact digital circuits.

We estimate the necessary area for a given circuit using the concept of gate equivalent. This is the basic unit of measure for digital circuit complexity [6]. It is based upon the number of logic gates that should be interconnected to perform the same input/output behaviour. This measure is more accurate that the simple number of gates [6]. The number of gate equivalent and an average propagation delay for each kind of gate are given in Table 7.1. The data were taken form [6]. Note that only 2-input gates NOT, AND, OR, XOR, NAND, NOR, XNOR and 2:1-MUX are allowed as the complemented input signals are available at no extra cost.

Table 7.1. Gates symbols, number of gate-equivalent

Name	Code	Gate-equivalent	delay
NOT	0	1	0.0625
AND	1	2	0.2090
OR	2	2	0.2160
XOR	3	3	0.2120
NAND	4	1	0.1300
NOR	5	1	0.1560
XNOR	6	3	0.2110
MUX	7	3	0.2120

The XOR gate is a CMOS basic gate that has the behaviour of sum of products $x\overline{y} + \overline{x}y$, wherein x and y are the input signals. However, a XOR gate is not implemented using 2 AND gates, 2 NOT gates and an OR gate. A 2:1-multipliexer MUX is also a CMOS basic gate and implements the sum of products $x\overline{s} + ys$ wherein x and y are the first and the second input signals and s is the control signal. It is clear that a XOR and MUX gates are of the same complexity [6].

Let us formalise the fitness function. For this purpose, let C be a digital circuit that uses a subset (or the complete set) of the gates given in Table 7.1. Let $gates(C)$ be a function that returns the set of all gates of C. On the other hand, let $val(T)$ be the Boolean value that C propagates for the input Boolean vector T assuming that the size of T coincides with the number of input signal required for C. The fitness function is given as follows, wherein X represents the input values of the input signals while Y represents the expected values of the output signals of C, n denotes the number of output signals that C has.

Let us formalise the fitness function. For this purpose, let C be a digital circuit that uses a subset (or the complete set) of the gates given in Table 7.1. Let $gates(C)$ be a function that returns the set of all gates of circuit C and $levels(C)$ be a function that returns the set of all the gates of C grouped by level. Notice that the number of levels of a circuit coincides with the cardinality of the set expected from function $levels$. On the other hand, let $val(X)$ be the Boolean value that the considered circuit C propagates for the input Boolean vector X assuming that the size of X coincides with the number of input signal required for circuit C. The fitness function, which allows us to determine how far an evolved circuit adheres to the specified constraints, is given in Equation 7.1, wherein X represents the values of the input signals while Y represents the expected values of the output signals of circuit C, n denotes the number of output signals that circuit C has, function $delay$ returns the propagation delay of a given gate as shown in Table 7.1 and ω_1 and ω_2 are the weighting coefficients that allow us to consider both hardware area and response time to evaluate the performance of an evolved circuit, with $\omega_1 + \omega_2 = 1$. For implementation issue, we minimized the fitness function below for different values of ω_1 and ω_2.

$$\begin{aligned} Fitness(C) = {} & \sum_{j=1}^{n} \left(\sum_{i | val(X_i) \neq Y_{i,j}} penalty \right) \\ & + \omega_1 \times \sum_{g \in gates(C)} gateEqui(g) \\ & + \omega_2 \times \sum_{l \in levels(C)} \max_{g \in l} delay(g) \end{aligned} \tag{7.1}$$

For instance, consider the evolved specification of Fig. 7.1. It should propagate the output signals of Table 7.2 that appear first (i.e. before symbol /) but it actually propagates the output signals that appear last (i.e. those after symbol /). Observe that signals Z_2 and Z_1 are correct for every possible combination of the input signals. However, signal Z_0 is correct only for the combinations 1010 and 1111 of the input signals and so for the remaining 14 combinations, Z_0 has a wrong value and so the circuit should be penalised 14 times. Applying function *gates* to this circuit should return 5 AND gates and 3 OR gates. Now applying function *levels* to this circuit should return two levels: {AND, AND, OR, OR, OR } and {AND, AND, AND }. If penalty is set to 10 then, function $Fitness$ should return $140 + 0.5(5 \times 2 + 3 \times 1) + 0.5(0.216 + 0.209)$. This fitness sums up to 21.7125. Note that for a correct circuit the first term in the definition of function $Fitness$ is zero and so the value returned by this

Table 7.2. Truth table of the circuit whose schematics are given in Fig. 7.1

X_1	X_0	Y_1	Y_0	Z_2	Z_1	Z_0
0	0	0	0	0/0	0/0	**0/1**
0	0	0	1	0/0	0/0	**0/1**
0	0	1	0	0/0	0/0	**0/1**
0	0	1	1	0/0	0/0	**0/1**
0	1	0	0	0/0	0/0	**0/1**
0	1	0	1	0/0	0/0	**0/1**
0	1	1	0	0/0	0/0	**1/0**
0	1	1	1	0/0	0/0	**1/0**
1	0	0	0	0/0	0/0	**0/1**
1	0	0	1	0/0	0/0	**1/0**
1	0	1	0	0/0	1/1	1/1
1	0	1	1	0/0	1/1	**0/1**
1	1	0	0	0/0	0/0	**0/1**
1	1	0	1	0/0	0/0	**0/1**
1	1	1	0	0/0	1/1	**1/0**
1	1	1	1	1/1	0/0	0/0

function weighted sum of the area required and the response time imposed of the evaluated circuit.

```
ENTITY C1 IS
  PORT (X: IN  BIT_VECTOR(1 DOWNTO 0);
        Y: IN  BIT_VECTOR(1 DOWNTO 0);
        Z: OUT BIT_VECTOR(2 DOWNTO 0));
END C1;
ARCHITECTURE Data_Flow OF C1 IS
   SIGNAL T: BIT_VECTOR(0 DOWNTO 0);
BEGIN
   T[0] <= X[1] AND Y[1];
   Z[2] <= (X[0] AND Y[0]) AND T[0];
   Z[1] <= T[0] AND (X[0] NAND Y[0]);
   Z[0] <= (X[1] NAND Y[0]) AND (X[0] NAND Y[1]);
END Data_Flow;
```

Fig. 7.1. VHDL data-flow specification for the circuit whose behaviour is given in Table 7.2

Apart from fitness evaluation, encoding of individuals is another of the important implementation decisions one has to take in order to use evolutionary computation in general and hardware design in particular. It depends highly on the nature of the problem to be solved. There are several representations that have been used with success: *binary encoding* which is the most common mainly because it was used in the first works on genetic algorithms, represents an individual as a string of bits; *permutation encoding* mainly used in ordering problem, encodes an individual as a sequence of integer; value encoding represents an individual as a sequence of values that are some evaluation of some aspect of the problem; and *tree encoding* represents an individual as tree. Generally, the tree coincides with the concrete tree as opposed to abstract tree of the computer program, considering the grammar of the programming language used. In the next sections, we investigate two different internal representation of digital circuits and look at the pros and cons of both of the methodologies.

7.3 Circuit Designs = Programs

In this first approach [4], a circuit design is considered as register transfer level specification. Each instruction in the specification is an output signal assignment. A signal is assigned the result of an expression wherein the operators are those listed in Table 7.1.

7.3.1 Encoding

We encode circuit specifications using an array of concrete trees corresponding to its signal assignments. The ith tree represents the evaluation tree or DAG (Directed Acyclic Graph) of the expression on the left-hand side of the ith signal assignment. Leaf nodes are labelled with a literal representing a single bit of an input signal while the others are labelled with an operand. A chromosome with respect to this encoding is shown in Fig. 7.2. Note that a (sub)expression can be shared.

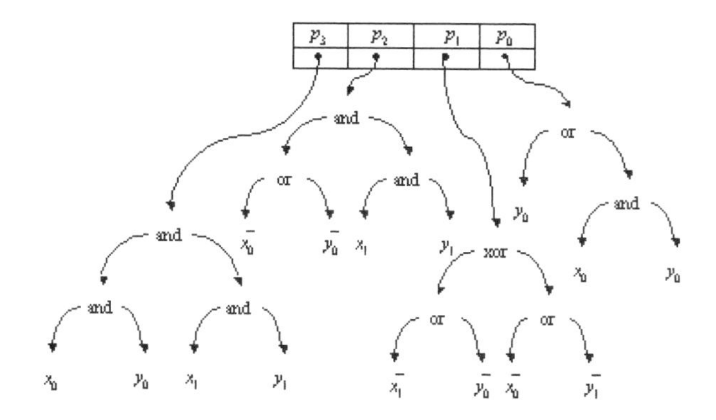

Fig. 7.2. Chromosome with respect to the first encoding

7.3.2 Genetic Operators

Crossover recombines two randomly selected individuals into two fresh offspring. It may be *single-point* or *double-point* or *uniform* crossover [3]. Crossover of circuit specification is implemented using single-point and double-point crossovers [3] as described in Fig. 7.3 and Fig. 7.4. We also analysed the impact of a variable double-point crossover that may degenerate into a single-point one when one of the points is 1 or coincides with the number of the expected output signals.

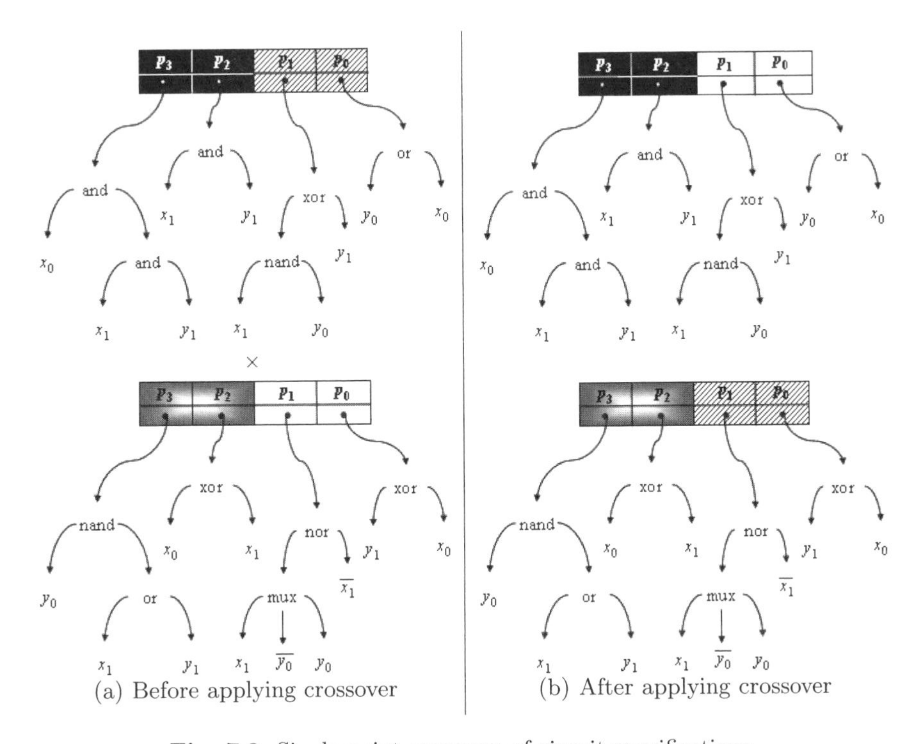

(a) Before applying crossover

(b) After applying crossover

Fig. 7.3. Single-point crossover of circuit specifications

The impact of each of the crossover techniques implemented is depicted in the chart of Fig. 7.5. The data used were obtained from 100 runs of the evolutionary process for the same input/output behaviour. The best runs in terms of convergence are represented. Fig. 7.5 shows clearly that the evolutionary process converges faster when the variable crossover is exploited. This is a crossover whose crossing points are generated randomly each time a crossover operation takes place.

We monitored the convergence point after 600 generations for the three crossover techniques and computed the relative gain occasioned. The results

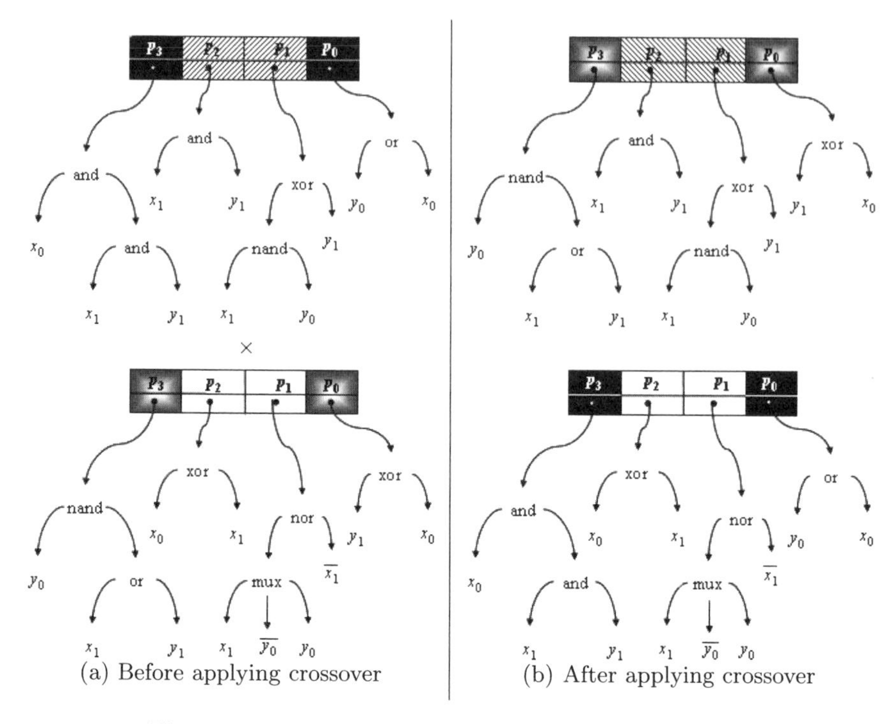

Fig. 7.4. Double-point crossover of circuit specifications

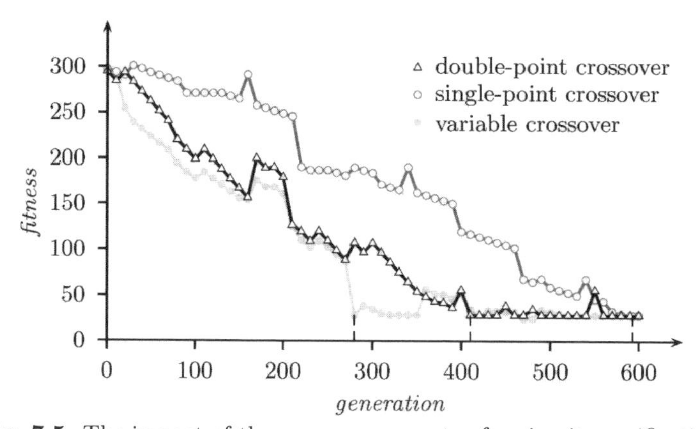

Fig. 7.5. The impact of the crossover operator for circuit specifications

are given in Table 7.3. The speedup is relative to the single-point crossover and is computed as the ratio of the number of generations when using the single-point crossover and that when using the crossover technique in question. The optimisation objectives, i.e. size and time were considered of equal preference and so $\omega_1 = \omega_2 = 0.5$. From the figures given in Table 7.3, we can notice that the variable crossover technique is the most efficient and should be used to speedup the evolutionary process of circuit design.

Table 7.3. Performance comparison of the specification crossover variations

Crossover	Convergence point	Gain
single-point	593	1.00
double-point	410	1.45
variable	280	2.12

One of the important and complicated operators for genetic programming [2] is the *mutation*. It consists of changing a gene of a selected individual. The number of individuals that should suffer mutation is defined by the *mutation rate* while how many genes should be altered within a chosen individual is given by the *mutation degree*.

Here, a gene is the expression tree on the left hand side of a signal assignment symbol. Altering an expression can be done in two different ways depending the node that was randomised and so must be mutated. A node represents either an operand or operator. In the former case, the operand, which is a bit in the input signal, is substituted with either another input signal or *simple* expression that includes a single operator as depicted in Fig. 7.6. The decision is random. In the case of mutating an operand node to an operator node, we proceed as Fig. 7.7.

7.4 Circuit Designs = Schematics

Instead of textual specification, a circuit design can also be represented by a graphical one, which is nothing but the corresponding schematics. So in this second approach [5], a circuit design is considered as map of gates given in Table 7.1. The first row of gates receives the input signals while the last row delivers the circuit output signals.

7.4.1 Encoding

We encode circuit schematics using a matrix of cells that may be interconnected. A cell may or may not involved in the circuit schematics. A cell consists of two inputs or three in the case of a MUX, a logical gate and a single output.

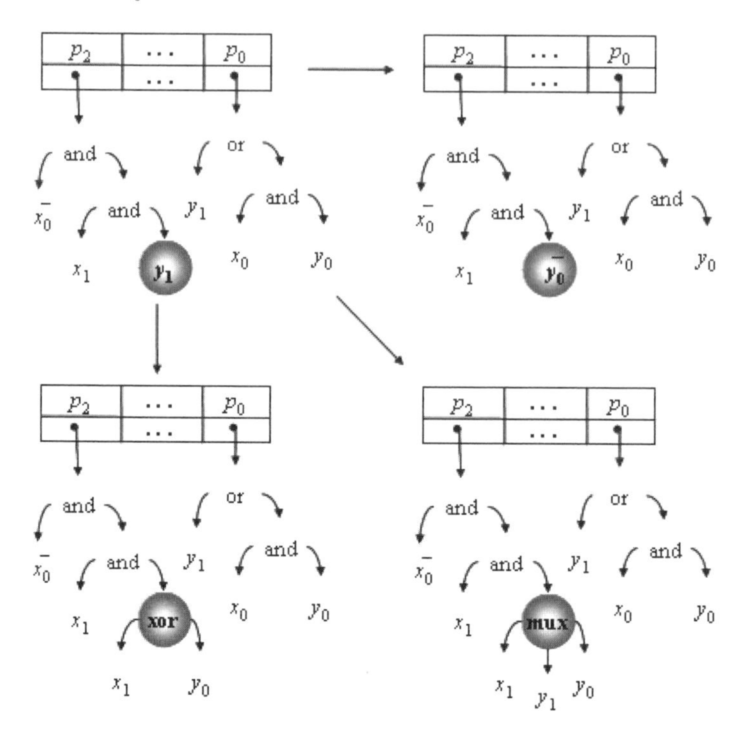

Fig. 7.6. Operand node mutation for circuit specification

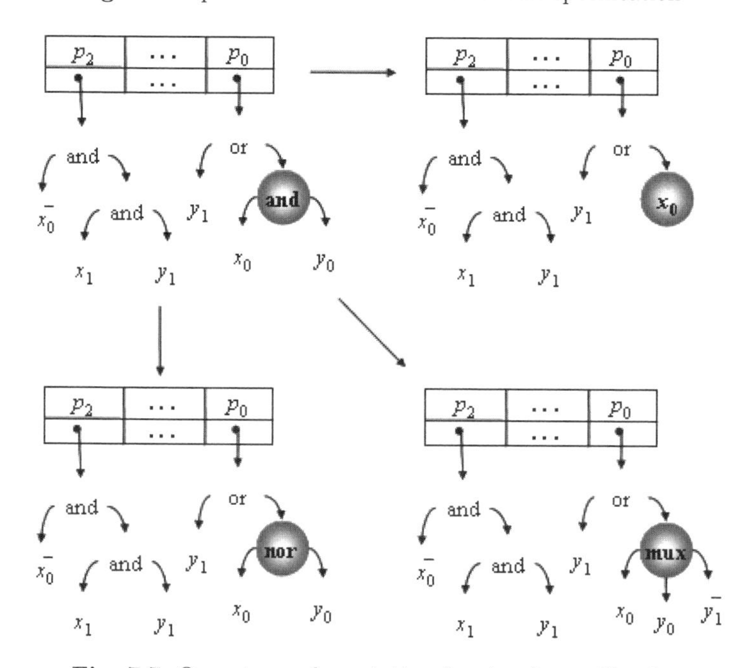

Fig. 7.7. Operator node mutation for circuit specification

A cell may draw its input signals from the output signals of gates of previous rows. The gates includes in the first row draw their inputs from the circuit global input signal or their complements. The circuit global output signals are the output signals of the gates in the last raw of the matrix. A chromosome with respect to this encoding is given in Fig. 7.8.

We encode circuit schematics using a matrix of cells that may be interconnected. A cell may or may not be involved in the circuit schematics. A cell consists of two inputs or three in the case of a MUX, a logical gate and a single output. A cell may draw its input signals from the output signals of gates of previous rows. The gates include in the first row draw their inputs from the circuit global input signal or their complements. The circuit global output signals are the output signals of the gates in the last raw of the matrix. A chromosome with respect to this encoding is given in Table 7.4. It represents the circuit of Fig. 7.8. Note that the input signals are numbered 0 to 3, their corresponding negated signals are numbered 4 to 7 and the output signals are numbered 16 to 19. If the circuit has n outputs with $n < 4$, then the signals numbered 16 to n are the actual output signals of the circuit.

Table 7.4. Chromosome for the circuit of Fig. 7.8

$\langle 1,0,2,8\rangle$	$\langle 5,10,9,12\rangle$	$\langle 7,13,14,11,16\rangle$
$\langle 2,4,3,9\rangle$	$\langle 1,8,10,13\rangle$	$\langle 3,11,12,17\rangle$
$\langle 3,1,6,10\rangle$	$\langle 4,9,8,14\rangle$	$\langle 7,15,14,15,18\rangle$
$\langle 7,5,7,7,11\rangle$	$\langle 4,10,11,15\rangle$	$\langle 1,11,15,19\rangle$

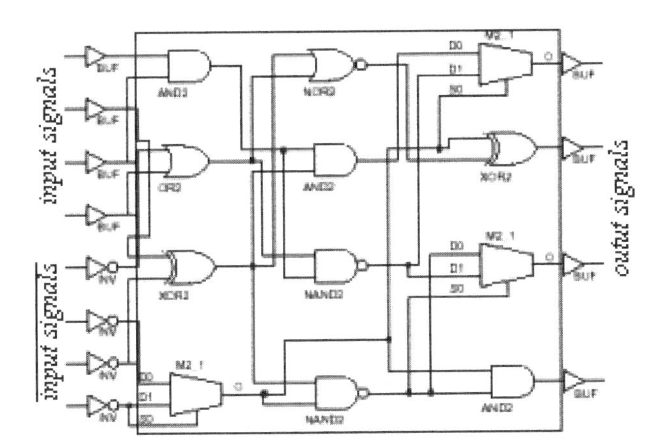

Fig. 7.8. Chromosome with respect to the second encoding

7.4.2 Genetic Operators

Crossover of circuit schematics, as for specification crossover, is implemented using four-point crossover. This is described in Fig. 7.9. The four-point crossover can degenerate to either the triple, double or single-point crossover. Moreover, These can be either *horizontal* or *vertical*.

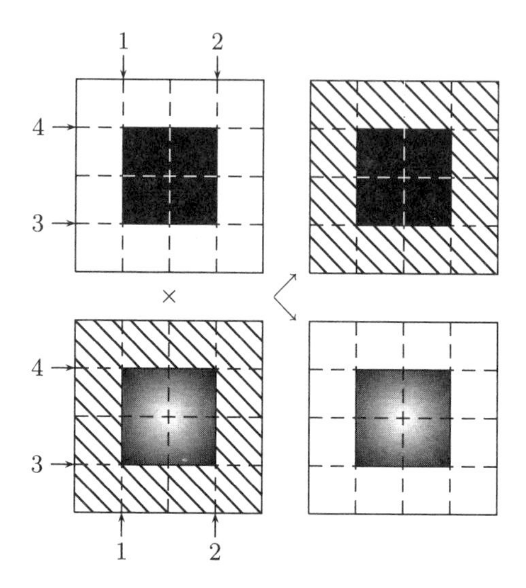

Fig. 7.9. Four-point crossover of circuit schematics

The four-point crossover degenerates into a triple-point one whenever one of the crossover points (say p) coincides with one of the sides of the circuit geometry. It is a horizontal triple-point when p is one of the vertical crossover points, as in Fig. 7.10–(a) and Fig. 7.10–(b) and horizontal otherwise, as in Fig. 7.10–(c) and Fig. 7.10–(d).

The four-point crossover degenerates into a double-point one whenever either both horizontal points or vertical points coincide with the low and high or left and right limits of the circuit geometry respectively. The horizontal double-point crossover is described in Fig. 7.11–(a) and the horizontal one in Fig. 7.11–(b).

The four-point crossover can also yield a horizontal and vertical single-point crossover as described in Fig. 7.12–(a) and Fig. 7.12–(b) respectively. The vertical single-point crossover happens when the horizontal crossover points coincide with the low and high limits of the circuit geometry and the one of the vertical points coincides with the left or right limit of the circuit geometry. On the other hand, the horizontal single-point crossover happens when the vertical crossover points coincide with the left and right limits of

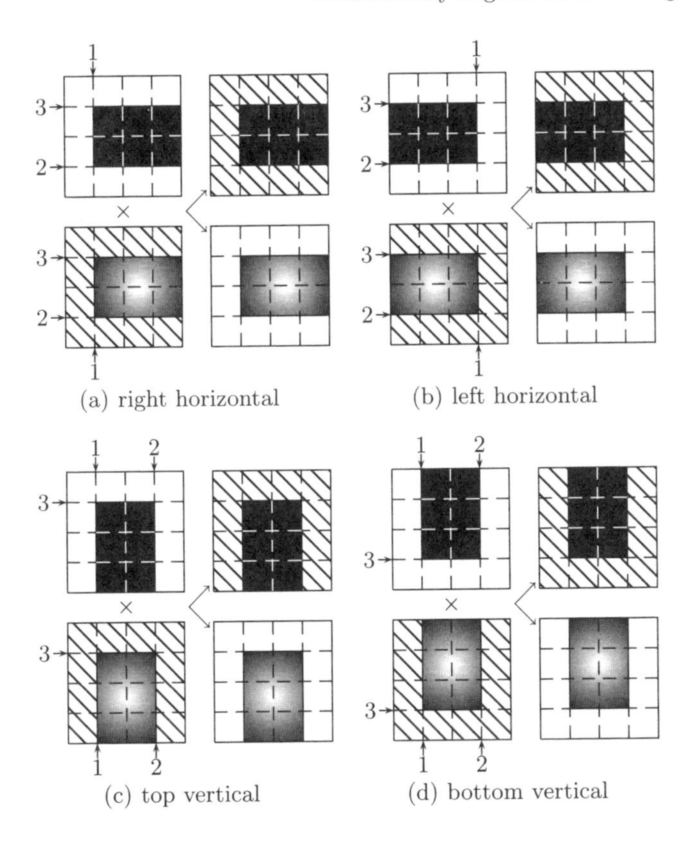

(a) right horizontal (b) left horizontal

(c) top vertical (d) bottom vertical

Fig. 7.10. Triple-point crossover of circuit schematics

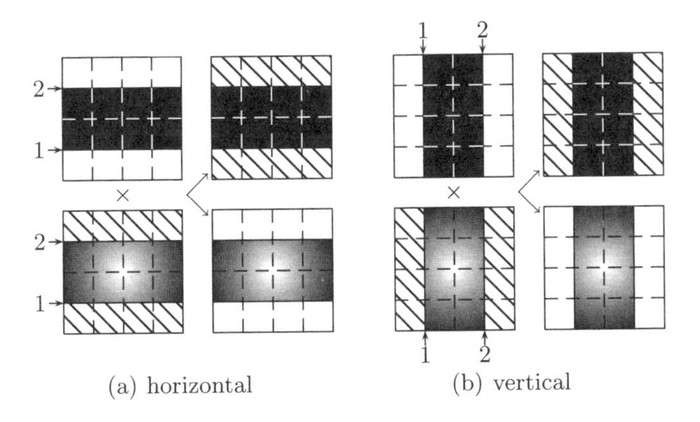

(a) horizontal (b) vertical

Fig. 7.11. Double-point crossover of circuit schematics

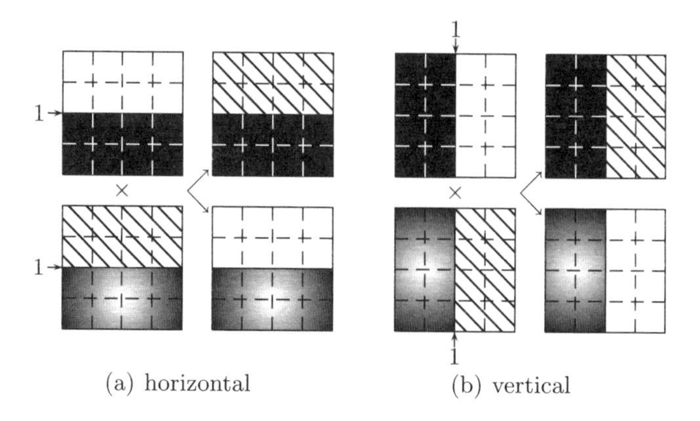

Fig. 7.12. Single-point crossover of circuit schematics

the circuit geometry and one of the horizontal crossover points coincide with low or high limit of the circuit geometry.

We studied the impact of each of the four-point, triple-point, double-point and single-point crossover. The purpose is to identify the best suited crossover technique, i.e. the one that allows a faster convergence to the minimal circuit design. The chart of Fig. 7.13 plots the average fitness function over 100 runs of the evolutionary process for the input/output behaviour that was used in the analysis of Fig. 7.5. This was done for each type of crossover. For all crossover techniques, except four-point crossover, we allowed variable crossing points. So, the triple-point crossover can be one of those depicted in Fig. 7.10, the double-point crossover can be one of those showed in Fig. 7.11 and finally, the single-point crossover can be one of those described in Fig. 7.12. The chart of Fig. 7.13 also plots the variable crossover which cab simply be one of the crossover operators described so far for circuit schematics. In this case, the four crossing points are generated randomly every time the operation takes place. So, the circuit schematics can be recombined according to one of the techniques showed in Fig. 7.9 – Fig. 7.12.

As before, we monitored the convergence point after 400 generations for the five crossover techniques and computed the speedup occasioned. The results are given in Table 7.5. Recall that the speedup is relative to the single-point crossover and is computed as the ratio of of generation number. Furthermore, the optimisation objectives, i.e. size and time were considered of equal importance. From the figures given in Table 7.5, we can establish that the variable crossover technique is the most efficient and should be used to speedup the evolutionary process of circuit design.

The mutation operator can act on two different levels: gate mutation or route mutation. In the first case, a cell is randomised and the corresponding gate changed. When a 2-input gate is altered by another 2-input gate, the

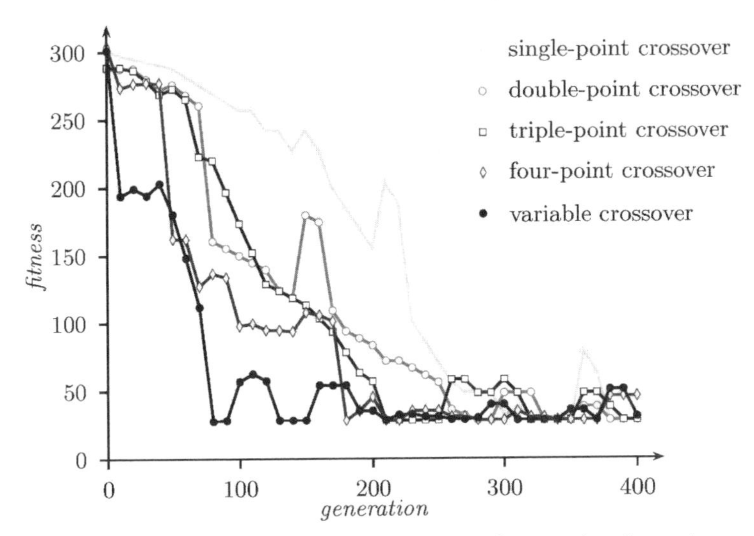

Fig. 7.13. The impact of the crossover operator for circuit schematics

Table 7.5. Performance comparison of the schematics crossover variations

Crossover	Convergence point	speedup
single-point	351	1.00
double-point	285	1.23
triple-point	213	1.65
four-point	180	1.95
variable-point	93	3.77

mutation is thus completed. However, when a 2-input gate is changed to a 3-input gate (i.e. to a MUX), the mutation operator randomise an additional signal among those allowed (i.e. all the input signals, their complements and all the output signals of the cells in the rows previous). Finally, when a MUX is mutated to a 2-input gate, the selection signal is simply eliminated. The second case, which consists of route mutation is quite simple. As before, a cell is randomised and one of its input signals is chosen randomly and mutated using another allowed signal. The mutation process is described in Fig. 7.14.

7.5 Result Comparison

In this section, we compare the evolutionary circuits yield when using the first encoding and those evolved using the second encoding. For this purpose, we use five examples that were first used in [1]. For each one of these examples, we provide the expected input/output behaviour of the circuit and the circuit

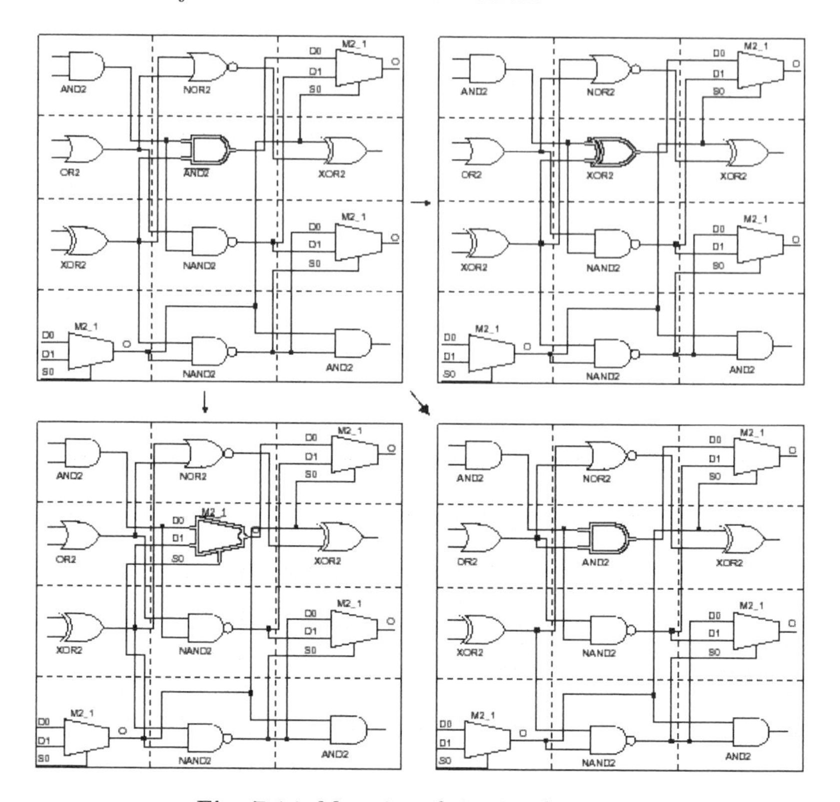

Fig. 7.14. Mutation of circuit schematics

designs evolved by the first and the second approach. For each of the evolution process, we state the number of generation for the same input parameters.

The truth table of the benchmarks is given in Table 7.6. Apart from benchmark (a), which has only three inputs, the remaining benchmarks, i.e. (b), (c) and (d), need four inputs. On the other hand, apart from benchmark (d), which yields three output signals, the rest of the benchmarks, i.e. (a), (b) and (c) propagate a single output signal. The circuits receive input signal X and propagate output signal Y.

The resulted circuit for the benchmark (a) using the first encoding is are given through the respective schematics in Fig. 7.15 as well as the corresponding VHDL data-flow specification in Fig. 7.16.

The circuit that was genetically programmed for the benchmark (b) (i.e., using the first encoding) is are shown by the corresponding schematics in Fig. 7.17 as well as the respective VHDL data-flow specification in Fig. 7.18.

The circuit evolved for the benchmark (c) by genetic programming is are shown by the corresponding schematics in Fig. 7.19 as well as the respective VHDL data-flow specification in Fig. 7.20.

Table 7.6. Truth table of the circuit used as benchmarks to compare both encoding methodologies

X_3	X_2	X_1	X_0	Y_a	Y_b	Y_c	Y_{d_2}	Y_{d_1}	Y_{d_0}
0	0	0	0	0	1	1	1	0	0
0	0	0	1	0	1	0	0	1	0
0	0	1	0	0	0	1	0	1	0
0	0	1	1	1	1	0	0	1	0
0	1	0	0	0	0	1	0	0	1
0	1	0	1	1	0	0	1	0	0
0	1	1	0	1	1	1	0	1	0
0	1	1	1	0	1	1	0	1	0
1	0	0	0	-	1	1	0	0	1
1	0	0	1	-	0	1	0	0	1
1	0	1	0	-	1	1	1	0	0
1	0	1	1	-	0	0	0	1	0
1	1	0	0	-	0	0	0	0	1
1	1	0	1	-	1	1	0	0	1
1	1	1	0	-	0	1	0	0	1
1	1	1	1	-	0	1	1	0	0

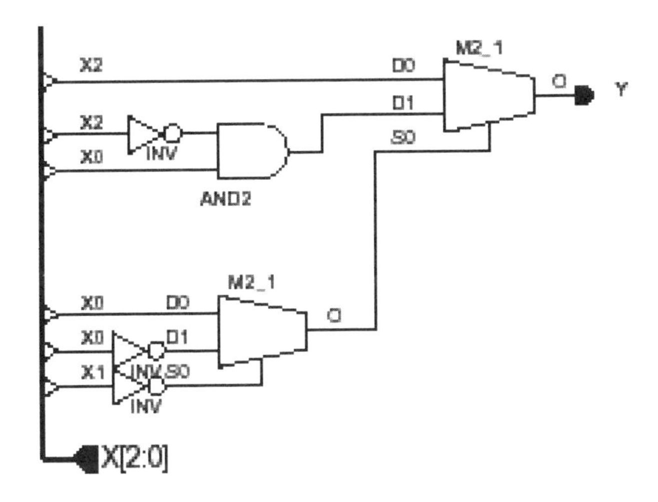

Fig. 7.15. Encoding 1: evolved circuit for the benchmark a

```
ENTITY C1_a IS
  PORT (X: IN  BIT_VECTOR(2 DOWNTO 0);
        Y: OUT BIT);
END C1_a;
ARCHITECTURE Data_Flow OF C1_a IS
BEGIN
   Y <= MUX(X[2], X[0] AND NOT X[2], MUX(X[0], NOT X[0], NOT X[1]));
END Data_Flow;
```

Fig. 7.16. Encoding 1: Data-Flow specification of the evolved circuit for the benchmark *a*

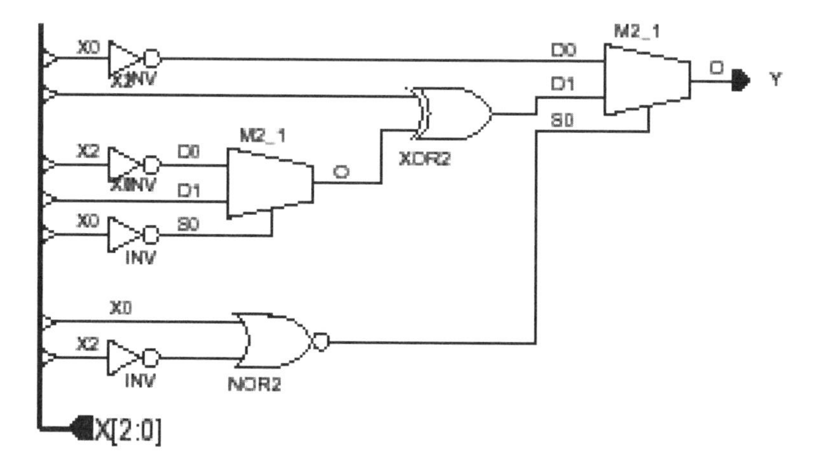

Fig. 7.17. Encoding 1: evolved circuit for the benchmark *b*

```
ENTITY C1_b IS
  PORT (X: IN  BIT_VECTOR(2 DOWNTO 0);
        Y: OUT BIT);
END C1_b;
ARCHITECTURE Data_Flow OF C1_b IS
   SIGNAL T: BIT_VECTOR(1 DOWNTO 0);
BEGIN
   T[0] <= NOT X[0];
   T[1] <= NOT X[2];
   Y    <= MUX(T[0], X[1] XOR MUX(T[1], X[1], T[0]), X[0] NOR T[1]));
END Data_Flow;
```

Fig. 7.18. Encoding 1: Data-flow specifcation of the evolved circuit for the benchmark *b*

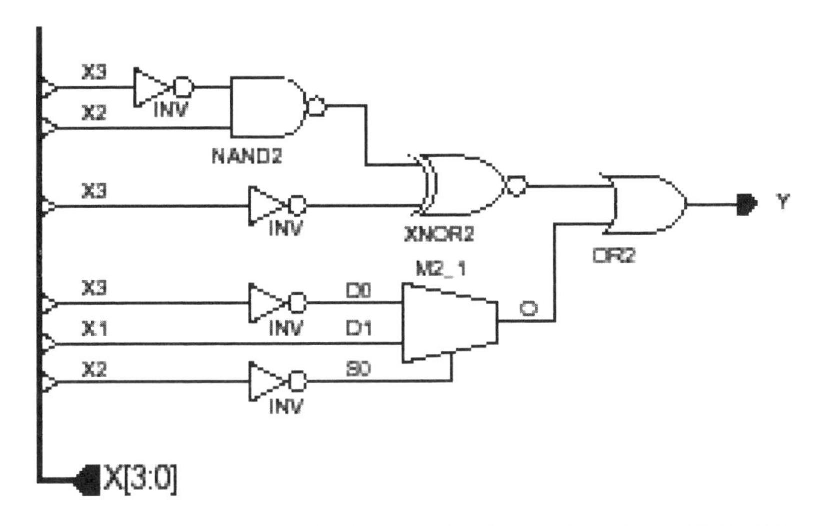

Fig. 7.19. Encoding 1: evolved circuit for the benchmark c in Table 7.6

```
ENTITY C1_c IS
  PORT (X: IN  BIT_VECTOR(3 DOWNTO 0);
        Y: OUT BIT);
END C1_c;
ARCHITECTURE Data_Flow OF C1_c IS
   SIGNAL T: BIT_VECTOR(0 DOWNTO 0);
BEGIN
   T[0] <= NOT X[3];
   Y    <= ((T[0] NAND X[2]) XNOR T[0]) OR MUX(T[0], X[1], T[0]);
END Data_Flow;
```

Fig. 7.20. Encoding 1: Data-flow specification of the evolved circuit for the benchmark c

The circuit evolved for the benchmark (d) by genetic programming is are shown by the corresponding schematics in Fig. 7.21 as well as the respective VHDL data-flow specification in Fig. 7.22.

Once again, for benchmark (a), we obtained the circuit whsoe schematics are given in Fig. 7.23 and the respective VHDL data-flow specification are shown in Fig. 7.24. This was done using the genetic algorithms (i.e., the second encoding).

The circuit evolved for the benchmark (b) GA (first encoding) is shown by the corresponding schematics in Fig. 7.25 as well as the equivalent data-flow specification in Fig. 7.26.

The circuit evolved for the benchmark (c) by genetic algorithm is are shown by the corresponding schematics in Fig. 7.27 as well as the respective VHDL data-flow specification in Fig. 7.28.

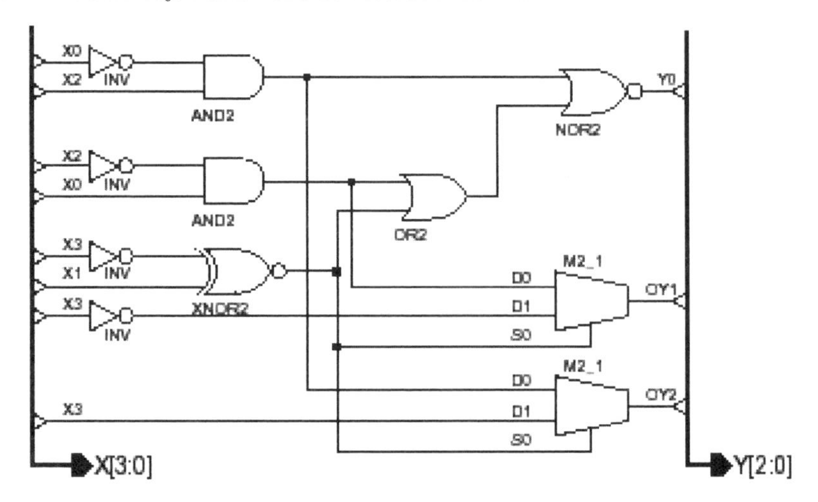

Fig. 7.21. Encoding 1: evolved circuit for the benchmark *d*

```
ENTITY C1_d IS
  PORT (X: IN  BIT_VECTOR(2 DOWNTO 0);
        Y: OUT BIT_VECTOR(2 DOWNTO 0));
END C1_d;
ARCHITECTURE Data_Flow OF C1_d IS
   SIGNAL T: BIT_VECTOR(3 DOWNTO 0);
BEGIN
   T[0] <= NOT X[3];
   T[1] <= T[0] NAND X[2];
   T[2] <= NOT X[2] AND X[0];
   T[3] <= T[0] XNOR X[1];
   Y[0] <= T[0] NOR (T[1] OR T[2]);
   Y[1] <= MUX(T[1], T[0], T[2]);
   Y[2] <= MUX(T[1], X[3], T[3]);
END Data_Flow;
```

Fig. 7.22. Encoding 1: Data-flow specification of the evolved circuit for the benchmark *d*

The circuit evolved for the benchmark (*d*) by genetic algorithm is are shown by the corresponding schematics in Fig. 7.29 as well as the respective VHDL data-flow specification in Fig. 7.29.

In Table 7.7, we provide the characteristics of the circuits evolved, which are the size in terms of gate-equivalent and the signal propagation delay in seconds. We also present the average numbers of generations that were necessary to reach the shown result. These numbers of generations were obtained for 100 different evolutions. In the last column of Table 7.7, we compute the average

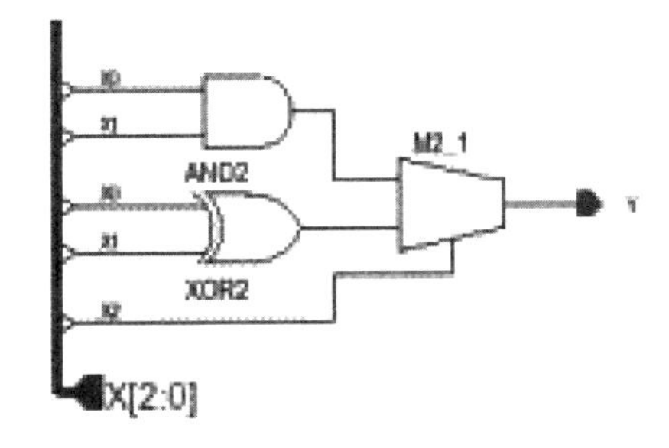

Fig. 7.23. Encoding 2: evolved circuit for the benchmark a in Table 7.6

```
ENTITY C2_a IS
  PORT (X: IN  BIT_VECTOR(2 DOWNTO 0);
        Y: OUT BIT);
END C2_a;
ARCHITECTURE Data_Flow OF C2_a IS
BEGIN
   Y <= MUX(X[0] AND X[1], X[0] XOR X[1], X[2]);
END Data_Flow;
```

Fig. 7.24. Encoding 2: Data-flow specification of the evolved circuit for the benchmark a

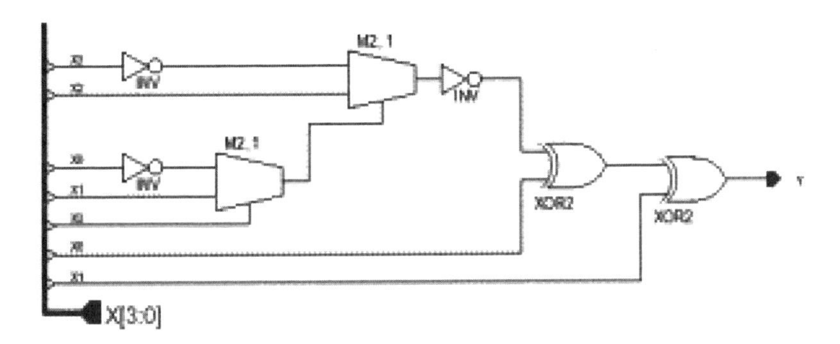

Fig. 7.25. Encoding 2: evolved circuit for the benchmark b in Table 7.6

speedup. Note that for both encodings, we exploited the variable crossover technique.

The graphical representation of the fitness values of the evolved circuits when using both encoding is depicted in Fig. 7.31 while that of the performance factor is given in Fig. 7.32. The numerical data are listed in Table 7.8. The hardware area and response time are considered to be of equal importance. The performance factor is computed as in (7.2) and the speedup is obtained using (7.3) which is simply the Amadahl's law.

```
ENTITY C2_b IS
  PORT (X: IN  BIT_VECTOR(2 DOWNTO 0);
        Y: OUT BIT);
END C2_b;
ARCHITECTURE Data_Flow OF C2_b IS
   SIGNAL T: BIT_VECTORO DOWNTO 0);
BEGIN
   T[0] <= NOT X[0];
   Y <= (NOT MUX(T[0], X[2], MUX(T[0], X[1], X[2])) XOR X[2]) XOR X[1];
END Data_Flow;
```

Fig. 7.26. Encoding 2: Data-flow specification of the evolved circuit for the benchmark *b*

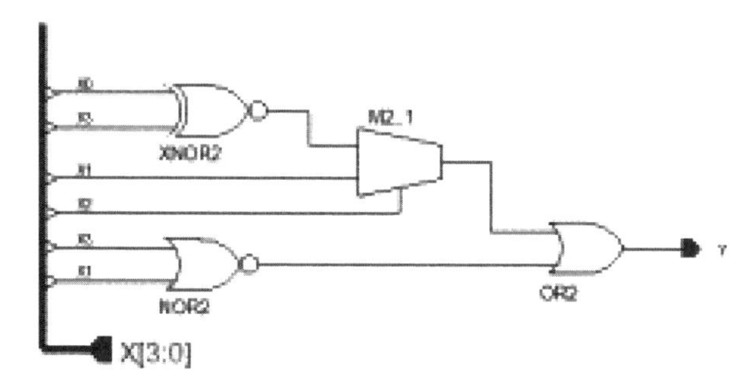

Fig. 7.27. Encoding 2: evolved circuit for the benchmark *c* in Table 7.6

```
ENTITY C2_c IS
  PORT (X: IN  BIT_VECTOR(3 DOWNTO 0);
        Y: OUT BIT);
END C2_c;
ARCHITECTURE Data_Flow OF C2_c IS
BEGIN
   Y <= (X[3] XNOR X[1]) OR MUX(X[0] XNOR X[3], X[1], X[2]);
END Data_Flow;
```

Fig. 7.28. Encoding 2: Data-flow specification of the evolved circuit for the benchmark *c*

$$performance = \frac{100 \times fitness}{\#generations} \tag{7.2}$$

$$speedup = \frac{performance_2}{performance_1} \tag{7.3}$$

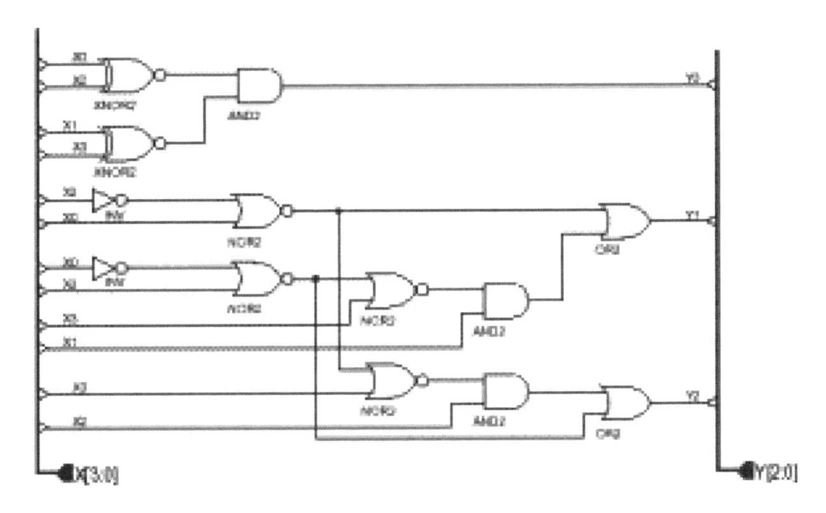

Fig. 7.29. Encoding 2: evolved circuit for the benchmark d in Table 7.6

```
ENTITY C2_d IS
  PORT (X: IN  BIT_VECTOR(2 DOWNTO 0);
        Y: OUT BIT);
END C2_d;
ARCHITECTURE Data_Flow OF C2_a IS
   SIGNAL T: BIT_VECTOR(1 DOWNTO 0);
BEGIN
   T[0] <= NOT X[2] NOR X[0];
   T[1] <= NOT X[0] NOR X[3];
   Y[0] <= (X[1] XNOR X[2]) AND (X[1] XOR X[3]);
   Y[1] <= T[0] OR ((T[1] NOR X[3]) AND X[1]);
   Y[2] <= T[1] OR (X[2] AND (T[0] NOR X[3]));
END Data_Flow;
```

Fig. 7.30. Encoding 2: Data-flow specification of the evolved circuit for the benchmark d

Table 7.7. Number of gate-equivalent and generations required to evolve the circuits presented

Benchmark	$Size_1$	$Size_2$	$Size_3$	$Time_1$	$Time_2$	$\#Gen_1$	$\#Gen_2$
(a)	11	8	12	0.4865	0.424	302	29
(b)	14	15	20	0.6985	0.973	915	72
(c)	12	9	20	0.6205	0.639	989	102
(d)	20	22	34	0.7015	0.888	1267	275

Table 7.8. Performance comparison of the impact of the studied encodings

Benchmark	$Fitness_1$	$Fitness_2$	$Performance_1$	$Performance2_1$	speedup
(a)	5.743	4.212	1.902	14.53	7.64
(b)	7.349	7.986	0.803	11.09	13.81
(c)	6.310	4.819	0.638	4.725	7.41
(d)	10.35	11.44	0.817	4.161	5.09

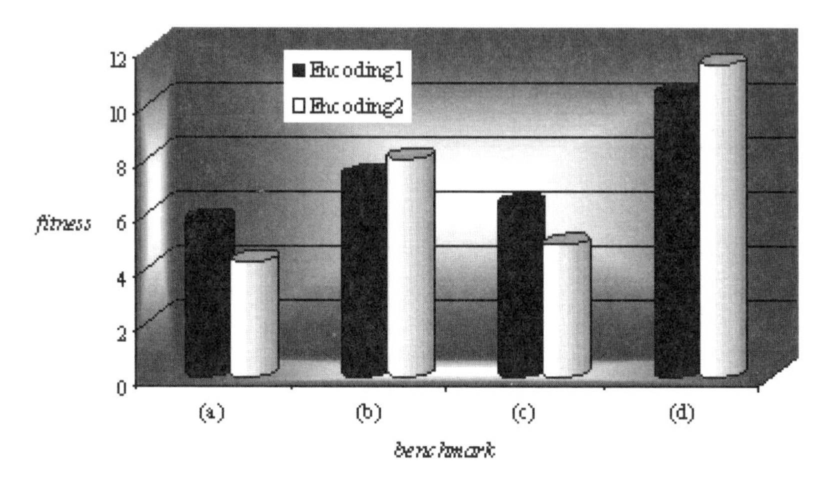

Fig. 7.31. Fitness comparison

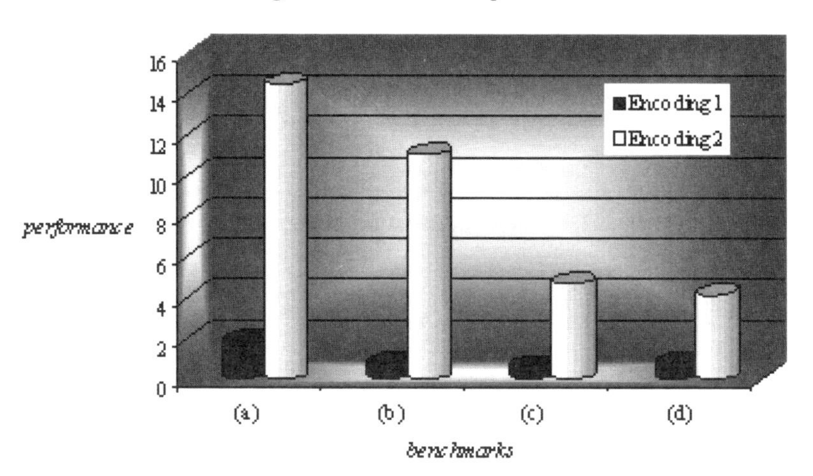

Fig. 7.32. Performance Comparison

7.6 Conclusion

In this chapter, we described two different encoding to be used in digital circuit evolution. In the first representation, we view a circuit design as a program that provide its specification. The second representation, in contrast, views a circuit as schematics that graphically describe the composition of the circuit. For each one of these encoding, we presented the variations of applicable genetic operators. We evaluated the impact of these variations on the convergence of the evolutionary design process. From the experiences performed with different crossover techniques, we can conclude that the variable crossover technique allows maintain a better diversity within the generational population and so yields an earlier convergence. Furthermore, we compared the performance of two genetic algorithms that use the described encoding. The comparison was peformed using 4 different benchmarks that were first used in [1]. The evolution process was repeated 100 times for each one of the benchmarks and the average number of generation that was required to reach a valid solution was recorded. In general, the evolved circuits are of the same complexity in terms of size and times. However, the number of generation necessary to reach a valid solution when using the first encoding was almost 10 times bigger. We computed the speedup according to Amadahl's law. This was about 8.5 in favour of the second encoding. The size of a solution designed by human using classical simplification methods is also presented. The evolved solutions are always better compared the ones classically designed.

References

1. Coelho, A.A.C., Christiansen, A.D. and Aguirre, A.H., Towards Automated Evolutionary Design of Combinational Circuits, Comput. Electr. Eng., 27, pp. 1–28, (2001)
2. Koza, J. R., Genetic Programming. MIT Press, (1992)
3. Miller, J.F., Thompson, P. and Fogarty, T.C., Designing Electronics Circuits Using Evolutionary Algorithms. Arithmetic Circuits: A Case Study, In Genetic Algorithms and Evolution Strategies in Engineering and Computer Science, Quagliarella et al. (eds.), Wiley Publisher (1997)
4. Nedjah, N. and Mourelle, L.M., Evolvable Hardware Using Genetic Programming, Proc. 4th International Conference Intelligent Data Engineering and Automated Learning - IDEAL 2003, Hong Kong, China, Lecture Notes in Computer Science, Vol. 2690, Springer-Verlag, (2003)
5. Poli, R. Efficient evolution of parallel binary multipliers and continuous symbolic regression expressions with sub-machine code GP, Technical Report CSRP-9819, University of Birmingham, School of Computer Science, December (1998)
6. Rhyne, V.T., Fundamentals of digital systems design, F.F. Kuo (Ed.) Prentice-Hall Electrical Engineering Series, (1973)
7. Thompson, A., Layzel, P. and Zebelum, R.S., Explorations in design space: unconventional design through artificial evolution, IEEE Transactions on Evolutionary Computations, 3(3):81–85, (1996)

8

Evolving Complex Robotic Behaviors Using Genetic Programming

Michael Botros

Department of Electrical Communication and Electronics,
Cairo University, Giza, Egypt.

Genetic programming (GP) applies the evolution model to computer programs. It searches for the best fit computer program through the use of biologically inspired operators such as reproduction, crossover and mutation. It has proven to be a powerful tool for automatically generating computer programs for problems that are too complex or time consuming to be programmed by hand. Example of these problems is writing control programs for autonomous robots to achieve non trivial tasks.

In this chapter, a *survey* of different methods for evolving complex robotic behaviors is presented. The methods represent two different approaches. The first approach introduces hierarchy into GP by using library of procedures or new primitive functions. The second approach use GP to evolve the building modules of robot controller hierarchy. Comments including practical issues of evolution as well as comparison between the two approaches are also presented.

The two approaches are presented through a group of experiments. All experiments used one or more Khepera robots which is a miniature mobile robot widely used in the experiments of evolutionary robotics.

8.1 Introducing Khepera Robot

Khepera is a miniature mobile robot that is widely used in laboratories and universities in conducting experiments aiming at developing new control algorithms for autonomous robots. It was developed by the Swiss Federal Institute of Technology and manufactured by K-team [1] [2]. Khepera robot is cylindrical in shape with a diameter of 55 mm and a height of 30 mm. Its weight is about 70 gm. Its small size and weight made it ideal robotic platform for experiments of control algorithms that could be carried out in small environments such as a desktop.

M. Botros: *Evolving Complex Robotic Behaviors Using Genetic Programming*, Studies in Computational Intelligence (SCI) **13**, 173–191 (2006)
www.springerlink.com

The robot is supported by two wheels; each wheel is controlled by a DC motor that can rotate in both directions. The variation of the velocities of the two wheels, magnitude and direction, will result in wide variety of resulting trajectories. For example if the two wheels rotate with equal speeds and in same direction, the robot will move in straight line, but if the two velocities are equal in magnitude but different in direction the robot will rotate around its axis.

Fig. 8.1. Miniature mobile robot Khepera (with permission of K-team)

The robot is equipped with eight infrared sensors. Six of the sensors are distributed on the front side of the robot while the other two are placed on its back. The exact position of the sensors is shown in Fig. 8.2. The same sensor hardware can act as both ambient light intensity sensor and proximity sensor.

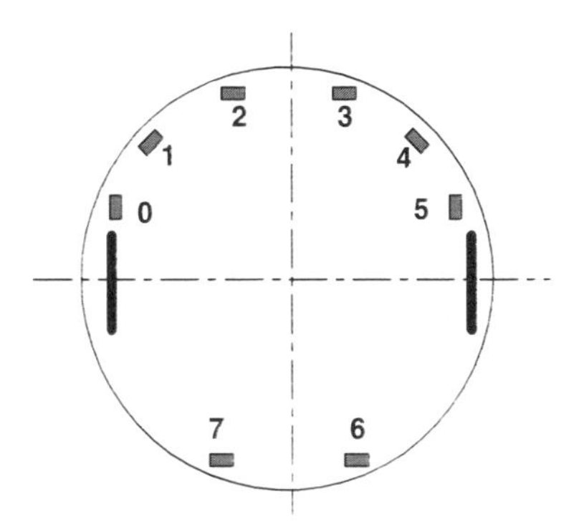

Fig. 8.2. The position of the eight sensors on the robot (with permission of K-team)

Each of the eight sensors consists of emitter and receiver parts so that these sensors can function as proximity sensors or ambient light sensors. To function as proximity sensors, it emits light and receive the reflected light intensity. The measured value is the difference between the received light intensity and the ambient light. This reading has range [0, 1023] and it gives a rough estimate how far the obstacles are. The higher reflected light intensity the closer obstacles are. It should be noted that we cannot find a direct mapping between the sensor reading and the distance from the obstacle, as this reading depends on factors other than the distance to the obstacle such as the color of the obstacle.

To function as ambient light sensors, sensors use only receiver part of the device to measure the ambient light intensity and return a value that falls in the range of [0, 1023]. Again, these measurements depend very strongly on many factors such as the distance to the light source and its direction.

An interesting feature of the Khepera robot is its autonomy, which includes autonomy of power and control algorithm. For the purpose of power autonomy, the robot is equipped with rechargeable batteries that can last for about 45 minutes. For experiments that may require much longer time, the robot can be connected to a host computer by a lightweight cable to provide it with the needed electrical power. This is an important feature that allowed long control experiments (such as developing evolutionary algorithms) to be carried out without repetitive recharging.

On the other hand, for the control autonomy, the robot's CPU board is equipped with MC68331 microcontroller with 512K bytes of ROM (system memory) and 256K bytes of RAM (user memory). This RAM memory can accommodate reasonable length program codes to provide control autonomy. The robot can be programmed using Cross-C compiler and the program will be uploaded to the robot through serial port communication with a host computer. Also the robot can be remotely controlled by a host computer where the control commands are sent to the robot through the serial link connection mentioned above. This mode of operation has an advantage of using computational power of the host computer.

8.2 Evolving Complex Behaviors by Introducing Hierarchy to GP

In this section, we will present how complex behaviors can be evolved by introducing hierarchy to genetic programming. Two different approaches will be presented. The first approach builds a subroutine library in GP. The subroutines in the library will be available to individual programs as an augmentation to the function set. The second approach uses function set that is more suitable to the nature of the input sensory information. The function set that will be presented here consists of neural networks. Each of the two different approaches will be presented along with an experiment that applied this approach to evolve a complex robotic behavior.

8.2.1 Genetic Programming with Subroutine Library

Providing a program solution to artificial intelligence problems (such as the problem of controlling a robot) usually includes building functions or procedures to do common calculations. They also discover the common similarities in the problem to be solved. The overall solution to the problem has the hierarchy of a main program that invoke a number of functions, which may in turn call other lower level functions. However, this program hierarchy is not available in the programs evolved using genetic programming. They consist of trees that make use of primitive functions without providing a technique to evolve procedures or functions along with the main program. Fig. 8.3 shows a typical program tree. The buildings units of this tree are the set of terminals T and the set of primitive functions F given by:

$$T = \{t_1, .., t_m\}$$
$$F = \{f_1, .., f_n\}$$

To overcome this problem in genetic programming, Koza in [3] [4] and [5] suggested a method for automatically detecting functions and subroutines and representing them inside the same tree of individual programs. In this case the program tree would consist of usual branches that form the program and return results, and a number of branches (l) each defining one function of the automatically defined functions ADFs. These ADF branches are not executed unless called by one of the value returning branches. Fig. 8.3 shows an example of program tree with ADF branches.

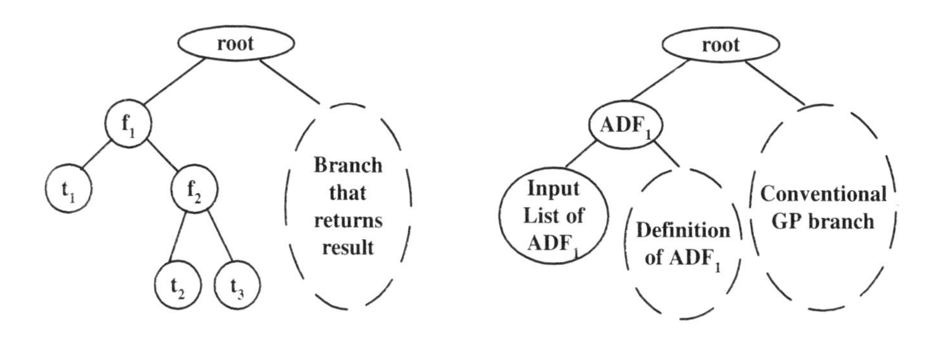

Fig. 8.3. Example of program tree in conventional GP and GP implementing ADF

The function set available to each individual is extended to include any of the l branches defining ADFs within the same individual tree. F in this case is given as follows:

$$F = \{f_1, .., f_n, ADF_1, .., ADF_l\}$$

But the ADF in a program subtree can not be used in another individual within the same generation, Furthermore, if an effective ADF is found as branch in a less fit individual, it will have small chance of reproduction and propagation to the next generations. These reasons led to thinking of a method to save well performing ADF functions in a subroutine library. In [6], Hondo et al. suggested a method for building that library and updating it according to fitness criteria. The function set in this case will be extended to include all the subroutines available in the library beside the ADFs represented as branches in the program tree as before. F in this case is given as follows:

$$F = \{f_1, .., f_n, ADF_1, .., ADF_l, SR_1, .., SR_L\}$$

where SR_i in this case is the subroutine number i in the library whose size is L.

The choice of a certain subroutine to be placed in the library depends on how well it performed when called by individual programs in the current generation. One possible fitness measure would be given by [6]:

$$fitness_{SR_i} = \frac{\sum_{j=1}^{C_i} f_j}{C_i}$$

where C_i is the number of individual programs that called subroutine SR_i and f_j is the fitness of each of these programs. By using fitness measures for subroutines, library is kept updated with effective subroutines at the expense of more computations.

The rest of this section will present a robotic experiment, by K. Yanai et al. [7], that maintained the subroutine library in the GP to evolve three robotic controllers for a team of three Khepera robots to simulate the escape behavior from a room during emergency time. The task set to these robots requires that they leave the room through an exit (marked with a color different from the color of the rest of the wall). To use this exit the robots have to press three buttons (given as black objects) in the room. This mission should be completed with a given limit of time. All the three robots are equipped with cameras to recognize different colors; however, each robot is allowed to move with a different speed. So the team in this case consists of similar robots but with different capabilities.

The genetic programming process used a different sub-population for each robot in the team. Individuals in each sub-population implemented ADF branches in addition to access to subroutine library in order to evolve hierarchical programs. The evolution process of each robot is not independent on the other two robots. Two aspects of interactions exist between the sub-populations. First, fitness evaluation of an individual from a given sub-population will take place while the best individuals in the other two sub-populations are used as controlling programs for the other two robots. Second, migration is performed among the three sub-populations to allow the exchange of possible individuals with high fitness.

The evolution process took place in simulated environments. Different simulated environments were used throughout the generations to evaluate the fitness of the individuals. Fig. 8.4 shows a possible example of these environments. The fitness functions used was the same for each individual in the three sub-populations. The fitness function, as seen in the next expression, encouraged early pushing of the three buttons (third term) and early leaving of the room (fourth term). The fitness function is given by [7]:

$$\begin{aligned} fitness = C_1 + C_2 &+ \text{bonus for pushing all buttons early} \\ &+ \text{bonus for leaving early} \end{aligned} \tag{8.1}$$

where C_1 is a constant added to the fitness for each robot robot that pushes a button and C_2 is a variable that is inversely proportional to the distance between each robot and the exit at the end of fitness evaluation.

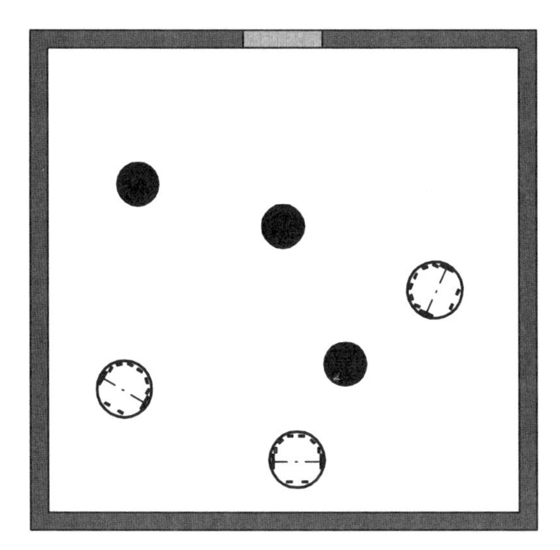

Fig. 8.4. Environment of the experiment

The evolution process lasted for 30 generations and the best fit individuals were tested in simulated as well as real environments. Evidence of coordination between robots was observed during these tests. Typical behavior of best fit individual in the environment can be described as follows: the fast robots rush to push the closest buttons, while the robot with the slowest speed would go directly to the exit since it can not afford to navigate the room searching for the buttons. The rest of the team would press any remaining buttons and then head to the exit [7].

This experiment showed possibility of evolving controlling programs that possess hierarchy similar to the hand-coded programs. This experiment and other robotics experiments applying hierarchical GP takes this approach one step higher by proving its validity in a more complex domain of problems. The next section will discuss another approach that utilizes neural networks as elements of the function sets.

8.2.2 Neural Networks and the Control of Mobile Robots

In the this section and the next one a method for combing the neural networks with genetic programming will be presented. The primitive function set was built using neural networks to detect different situations faced by the robot. This section will give a brief review of artificial neural networks (ANN) basics that will be useful for the reader who is unfamiliar with ANN for better understanding of next section.

Single neuron model: ANN consists of interconnection of processing units called neurons which is inspired by the biological neurons of the nervous system. In the biological neuron, it receives all inputs stimuli in the form of spiking electrical activity and these inputs are large enough they can activate the neuron so that it will in turn send electrical signals to other neurons connected to it. In our ANN model, the neuron receives the inputs $x_1, ..., x_n$ and sum them after being weighted by corresponding weights $w_1, ..., w_n$ (called synaptic weights) and if the result is above a threshold b then the net result can generate an output y through a nonlinear function Φ. The following equation describes the input-output relation of a class of artificial neurons. This single neuron model is usually represented graphically as shown in Fig. 8.5.

$$y = \phi(\sum_{i=1}^{n} w_i x_i - b)$$

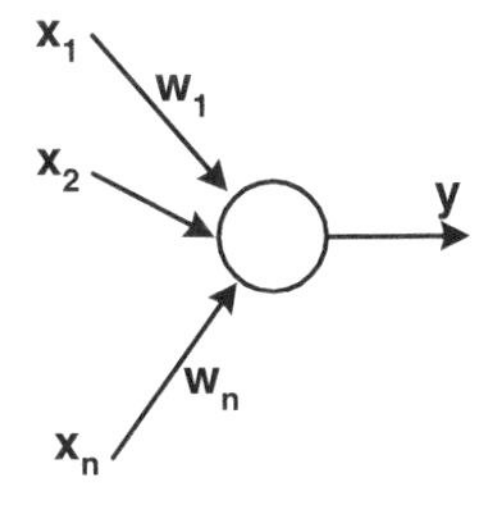

Fig. 8.5. Model of a single neuron

Interconnecting neurons: A widely used class of neural network architecture is obtained by arranging neurons in cascaded layers as shown in Fig. 8.6. Each layer consists of an array of neurons whose outputs are connected to the next layer inputs. The first layer inputs are the ANN inputs and the last layer outputs are the ANN outputs. Fig. 8.6 shows ANN with three layers: input layer, hidden layer, and output layer.

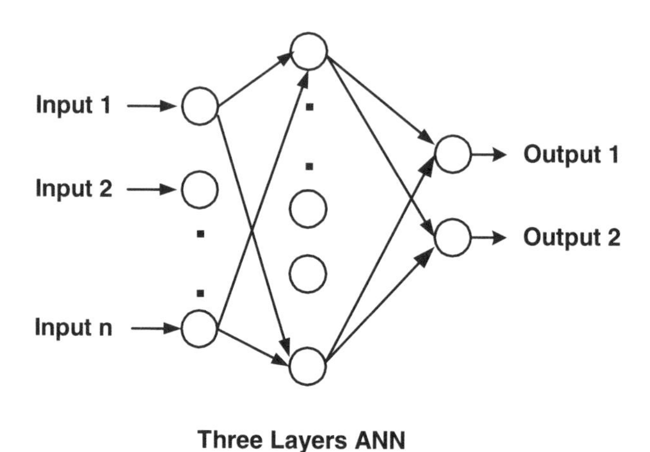

Fig. 8.6. Artificial neural network with three layers

Learning: Now we have an ANN with given architecture, how can it perform a given function or how can it give a desired output for a given input? The answer of this question is through the learning process of the neural network which includes adjusting the synaptic weights w's and threshold b of each neuron such that the overall ANN will give the desired output for a given input. A widely used, though not the only one, algorithm for learning the weights and thresholds is through back propagation. In this algorithm, pairs of inputs and corresponding desired outputs are used to learn the weights iteratively. In each iteration, the inputs are applied to the ANN and the error between the actual and desired output is used to update the weights in way to minimize the squared error. We can look at this type of learning as an adaptive algorithm or optimization problem in which the weights are optimized iteratively to minimize an objective function that depends on the error.

Why ANN is useful to robotics: The ability to lean by example is a desirable feature in neural networks that can benefit robotics. Examples of input sensory information and corresponding robot output (such as robot velocity in case of mobile robots) can be used to train neural networks to be employed as robotic controller. Another desirable feature is the highly nonlinear relation between the ANN inputs and outputs due to the nonlinear function ϕ of each neuron. This nonlinear relation is useful when we want to have an output with some degree of insensitivity to inputs corrupted with noise such as robot sensory inputs.

Evolutionary algorithms and neural networks: Genetic Algorithm GA can be incorporated with ANN in different ways. One way is to use the evolution process of the genetic algorithm to search for the optimal set of weights and thresholds. Example of this approach applied to robotics can be found in [9], in which GA acted as a optimization tool for the weights of a neural network controller. Another way is let the GA evolve the architecture of the ANN along with its weights. Example of this method applied to robotic application can be found in [10]. The next section discusses further incorporation between the genetic programming and ANN.

In this section, a brief introduction to ANN concepts was presented. The interested reader can find detailed treatment of ANN in dedicated textbooks such as [11].

8.2.3 Using Neural Networks as Elements of the Function Set

If we want to write a program to identify possible situations faced by the robot, it may be difficult to identify these situations using logic expressions in the sensors. For example, if we want to identify the situation in which there is an obstacle in front of the robot, a possible logic expression can be (sum of front sensors > numerical threshold). Given the noisy nature of real sensors, this expression may not be triggered in every time there is an obstacle in front of the sensors. Noise or distance to obstacle may cause the sensors values to be active but less than the given threshold. For the case of Khepera robot, and other robots utilizing infrared sensors, the proximity sensor readings depends on the color of the obstacle too as we mentioned earlier in the first section. One potential solution to this problem is using a neural network whose input comes from the proximity sensors (and possibly from ambient light sensors too to cover the cases of different obstacles with different colors) and whose output is used to identify these situations. The synaptic weights will need to be trained first using example situations to be identified. The remaining part of this section will present an experiment that used trained neural networks as primitive function in the genetic programming process.

The experiment is performed by K. Lee et al. [12]. The task of the robot is to reach a target location which is marked by the presence of a light source and at same time robot has to avoid the obstacles present in its environment. To reach its target, it will need the ambient light sensors I_0,.., I_7 to detect the light source and it will use the proximity sensors S_0,.., S_7 to avoid the obstacles.

During the robot path to the target it will encounter situations when the obstacles are in front of it or to its left or right. Also, avoiding obstacles may change its orientation with respect to the target so it will need to detect such situations (target to left, right, .., etc). The above described situation were chosen to be the function set of the GP while the terminal set was the basic action that can be taken by the robot such as moving forward, or backward and turning left or right. Typical code example would consist of a tree of elements of the function and terminal set such as the one depicted in Fig. 8.3 part (a) and as shown in figure 1 in [12].

Each of these situation described above was build as a function of the eight proximity sensors and eight ambient light sensors. These functions were implemented as a neural network that consisted of 16 input neurons and one output neuron in a fashion similar to the general layered architecture shown in Fig. 8.6. The synaptic weights were trained by back propagation. In back propagation the weights are adjusted iteratively to give the desired output. The error between the current and desired output is used to tune the weights as discussed in the previous section. Now, having trained the weights of the neural networks to detect the situations that can face the robot, they are ready to act as function set.

The genetic programming process took place in simulation. Different simulated environments were used for fitness evaluation. They contained arrangements of obstacles and two light sources. The robots will be rewarded for each time they reach any of light sources and would be punished for colliding with any obstacle. The fitness function used is given by [12]:

$$fitness = C_1 \frac{N_{collision}}{N_{steps}} + \frac{N_{steps} - C_2 N_{reaching\ target}}{N_{steps}} \tag{8.2}$$

where N_{steps} is the number of the steps the individual program is allowed to perform in each fitness evaluation, $N_{collisions}$ is the numbers of the collisions that occurred during these steps and $N_{reachingtarget}$ is the number of successful achieving of the task by approaching the light sources. The two constants C_1 and C_2 constitute a tradeoff between the importance of the avoiding obstacles and approaching the target.

After 100 generations, the results of the experiments showed the decrease of the average fitness and increase of the number of reaching the target with the increase of the number of generations. The best fit individual was transformed to a real robot and was tested in a real environment with obstacles and single light source. The robot was able to find its way between the obstacles to reach the light source in the corner of the room [12].

8.2.4 Comments

Two methods for including hierarchy inside the GP have been presented. First method evolve the main program along with the procedures and maintain a procedure library while the second method uses a new function set consisting of neural networks. In this section, we will discuss some of practical points in both methods.

- **Reducing human designer early choices in evolution process.** In the first method and in conventional GP too, human designer has to choose a set of primitive functions, their parameters, and write their code. On the other hand, the second method offers a more automated method for these choices. The learning process of neural network will tune the synaptic weights of the network to perform the right task by choosing the right sensors. For example, in the previous experiment neural networks were used to detect two sets of situations, the first set is related to the obstacle (obstacle to the left, to the right ... etc) and the other related to the target (target ahead, target to the left .. etc). After training, we expect that the weights associated with proximity sensors are higher than those corresponding to light sensors in the first set and less effective in determining the output of the network in the second set.

- **Moving from simulation to real robots.** Very few evolution experiments are carried on real robots in real environments. The common approach is evolving the controller in simulation and then testing the best fit individual using a real robot and real environment. This approach avoids lengthy experiment time and hardware problems if real robots were used. However, a drawback of this approach is difference between the robot behavior in simulated and real environments which may arise due to the noise present in real sensors which have not been accounted for in the simulation.
The hand coded primitive functions of first experiment had to be modified after transition from simulation to real worlds. For example in the first experiment of the robot team escape, function sets that detect colors were used (buttons and exit were differentiated based on their colors). These functions have to be edited to account for the difference between the simulated camera and the real camera [7]. In the second experiment too, the neurons threshold values had to be modified when testing the best fit individual on real robot [12]. However this step was necessary because the learning of the neural network was based on samples of the simulated sensor values. We believe that this modification could have been avoided if the learning process used samples of the real sensor data.

- **Size of the evolved programs.** First method of using hierarchical GP is associated with increasing the number of branches to implement ADF. Some methods were proposed to control the growth of the code [8]. For example, segments of code that will not be reached and will not be executed, called introns, can be discovered and removed. If I is a logical condition, then if the tree contains the expression (If true or I), I will be infective intron. GP can detect such introns by associating a counter with elements of terminal sets and functions and incrementing it each time they are evaluated. The expression whose all terminals and functions counters were not incremented could be a possible intron. Upon detecting an intron, GP can edit it to become effective or delete it to control the tree growth.

8.3 Evolving Complex Behaviors by Introducing Hierarchy to the Controller

In this section we will present two examples of evolving complex robotic behaviors by using GP to evolve the building modules of a hierarchical robot controller. The concept of using hierarchy in robot controllers is similar to concept of "divide and conquer" in software design and programming and it also follows the main design steps which are:

Step one: The complex behavior is broken down into simpler behaviors or tasks.
Step two: A program or controller is designed to implement this behavior.
Step three: A method is used for coordination or arbitration between different behaviors.

Using different methods to carry out steps one and three results in different architectures of controllers [13]. In the next subsections we will present two main architectures: subsumption architecture and action selection architecture which will both be used in the two examples that will come next.

8.3.1 Subsumption Architecture

In subsumption architecture [13], step one is carried by assigning each one of the simpler behaviors to one of the layers of the architecture with higher priority behaviors placed in upper layers. The sensory information is input to all layers, also outputs of the lower layers may be used as inputs to higher layers. For coordinating between the outputs of different layers, a fixed priority system is used which is predefined by the designer of the controller. The higher layers output has higher priority to propagate to the physical output of the controller such as the robot motor speeds. It can also inhibit or pass the output of lower layers to the controller output as shown in Fig. 8.7.

A simple example of subsumption architecture is a controller to enable the robot to explore its environment while avoiding the obstacles. This behavior could be divided into to basic behaviors: exploring the environment and avoiding the obstacle while the last one given higher priority to avoid damaging the robot or hurting humans in the environment.

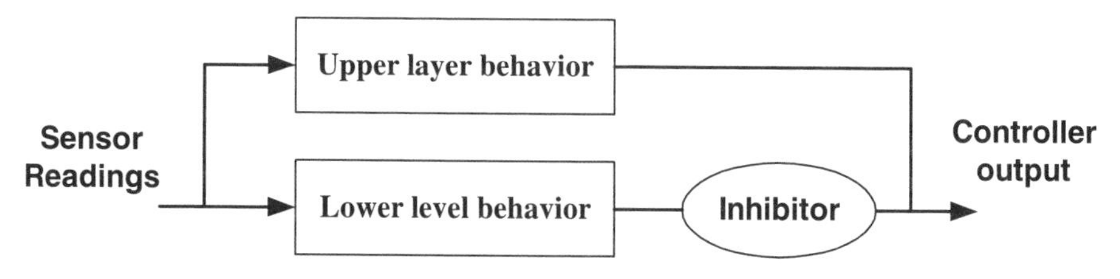

Fig. 8.7. Subsumption architecture

An apparent advantage of this architecture is simplicity in arbitrating between behaviors. Also, the behavior of the robot is expected to be dominated by the upper layer behaviors so we are able to predict the robot behavior in different environment. However some tasks may require the robot to switch between behaviors dynamically according to the current situation, for example searching for a specific object, then carrying it to a given target. In this case the first behavior (searching) should be dominant for sometime and then the robot needs to switch to the second behavior (carrying object) when the object is found. In this class of tasks, an architecture that handles dynamic priorities, such as action selection architecture, is needed.

8.3.2 Action Selection Architecture

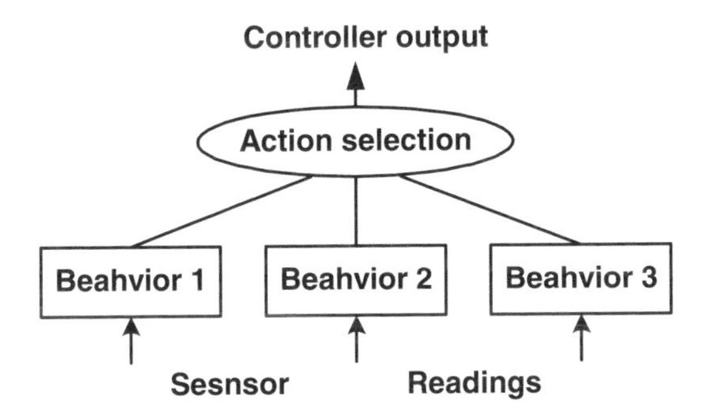

Fig. 8.8. Action selection architecture

In action selection [17], step one is done by breaking the complex behavior into simpler behaviors and associating with each of them an activation level which reflects its current importance and suitability to the situation. The activation level is function of the current robot's goal and its sensory readings. Step three is carried as follows: before each action taken by the robot, that output of each simple behavior module is evaluated as well as its activation level. The behavior with the highest activation level is allowed to propagate its output to the controller output. Since these activation levels are dynamically changing, action selection belongs to the class of architectures that use dynamic priorities.

Fig. 8.8 shows an example of action selection control architecture in which three behaviors compete, through their activation levels, to gain control over the robot. It should be noted that it is possible to have a tree-like structure within the action selection controller as shown in Fig. 8.8. In this case both the left and the right behaviors are further broken into even simpler behaviors. We will encounter another example of this tree-like architecture in one of the experiments presented after this subsection.

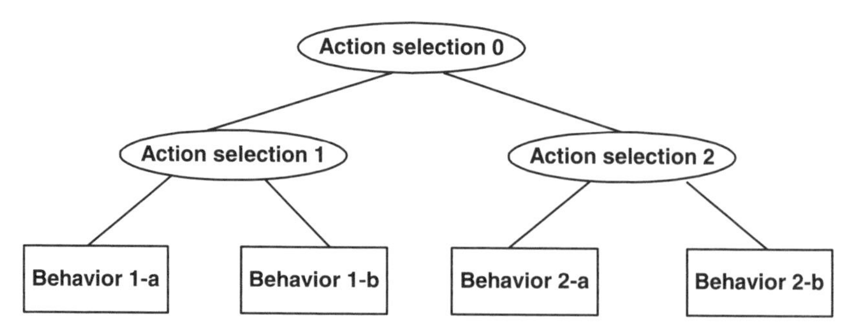

Fig. 8.9. Action selection architecture with tree-like structure

Having introduced these two architectures, we are now ready to move to the next section to see how GP can be used within each architecture to evolve its constituting modules.

8.3.3 Using GP to Evolve Modules of Subsumption Architecture

In this experiment, performed by S. Kamio et al. in [16], genetic programming is used to evolve the modules of a robot controller with a subsumption architecture. The controller is intended to act as central controller for a robot team consisting of two Khepera robots with different capabilities described as follows:

1. Robot 1 is a Khepera robot provided with a camera mounted on the top of the robot.
2. Robot 2 is a Khepera robot provided with a gripper module that is also mounted on the top of the robot. The gripper can perform two main actions: picking and releasing the object. The robot can detect the presence of an object in the gripper by using a light barrier sensor placed inside the gripper. Fig. 8.10 shows the robot while using the gripper to hold an object.

The task of the controller is to construct coordination between the robots so that the second robot, which is equipped with the gripper only without any vision capabilities, will locate objects found in the experiment environment and carry them to a certain goal area (marked by its different color). Clearly this would require a help from the first robot with the camera that would be used to estimate the distance between itself and the object (d_{object}) and the second robot (d_{robot2}) [16].

To achieve the previously described task, the central controller was designed using subsumption architectures, in which each subtask (or behavior) is assigned to one layer of the architecture with the higher layers able to process the outputs of lower layers and to decide whether to inhibit or pass it to the final outputs. These layers are described as follows [16]:

Fig. 8.10. Khepera robot with the gripper mounted on the top of it (with permission of K-team)

1. Lower layers: They are responsible for signal processing of camera signals and estimating d_{robot2} and d_{object}.

2. Intermediate layer: The task of this layer is to generate the motor speeds of the two robots based on the distance estimate from the previous layer.

   GP was used to evolve this module. Beside the inputs from the previous level, the terminal set included basic actions for the two robots such as moving forward and rotation to left and right. Generations consisted of 2000 individuals each. Each individual was tested in a number of simulated environments where this module was allowed to control the two robots for a number of steps till robot 2, with the gripper, has moved a distance equals twice the its initial distance to the object. After being tested in different environments, the fitness of the individual was evaluated as the maximum, over all testing environments, of the final distance between the object and the robot with gripper. After the end of one evaluation generation, the individuals of the new generation would then be produced by the crossover and mutation operator. The results of best evolved individual for this module will be presented after discussing the last layer.

3. Higher layer: The task of this layer is to decide when robot 2 would open its gripper to hold the object and when to release it. This layer was programmed by human programmers.

When the best fit individuals after the evolution process, the two robots showed the desired guided navigation behavior. Robot 1, with the camera, starts moving first acting like the explorer followed with robot 2. Compared to individuals of first generations, individuals in last generations showed more direct pathes to the object and to the goal area [16].

This experiment was example of using GP to evolve the modules of controller architecture. A question may rise here asking if it could be possible to evolve a program for the entire controller directly in a single GP process instead of evolving each module. This could be a possible approach, however in this case the genetic programming will search a larger space of programs that achieve the overall task rather than the smaller space of each module. It may seem that this single GP will save effort by designing only one fitness function and running the GP once, however it may need larger number of generations for convergence.

8.3.4 Using GP to Evolve Modules of Action Selection Architecture

In this experiment, performed by W. Lee et al. in [18], GP is utilized to evolve a robotic controller with action selection architecture. The task of the controller is to enable to robot to navigate in the environment to locate a cylindrical object and push it towards a target marked by its light source. Proximity sensors of the robot will help it locate the object while the ambient light sensors will help it locate the light source.

The complex task mentioned above can be decomposed using action selection architecture. At any time, the main controller will act as an arbitration or action selection module that arbitrates between two basic behaviors: (1) navigation in the environment and (2) pushing the object. The later can be further implemented as another action selection module that arbitrates between two sub-behaviors: (2-a) pushing the box forward and (2-b) rotating around the object, which will be useful when the object slips from the robot and it needs to adjust itself to a suitable pushing position relative to the object. This architecture can be implemented as a tree as shown in Fig. 8.9.

All the behaviors and action selection modules were evolved using GP. We will start first by the GP process of behavior modules then the action selection modules.

The behavior modules were represented as logic building blocks. The program will be built up using "And" and "Or" functions and comparators that compare the input sensor values with numerical constants. So the function set consisted of these comparators beside the logic gates while the terminal set consisted of the input sensors and the numerical constants [18].

The fitness function of behavior module was designed on the sensor level to encourage the individuals to acquire the desired behavior. For example, the behavior (2-a) of pushing the object forward encouraged high front sensor values so that the robot would maintain the object in its front direction.

The action selection modules are program units that select which behaviors to be executed. It can be thought of as a program that can output either "1" or "0" depending on the behavior selected, so by this way they can be built up using logic components as the behavior modules so the function set again consisted of logic gates. Since the selection of the suitable behavior will depend on the current situation which need to estimated using the robot sensors, the terminal set need to contain the include the input sensors. The readings of the input sensors will be compared with numerical constants to produce "1" or "0" results suitable to be used with the logic building blocks.

Evolution of each module was carried in simulated environments then tested in real environments. The overall controller, built using best fit individuals of each module, enabled the robot to navigate in the environment till finding the object, and then it switches to the other behavior of pushing the object [18].

8.4 Comments

Two different approaches for evolving complex behaviors were discussed. The first one introduces the hierarchy to the GP by allowing it to evolve functions along with the main program. The second one uses the hierarchy of the controller to evolve the building modules of the controller architecture.

- **Evolved program hierarchy.** In the first approach the hierarchy is automatically generated by the GP through the use of ADFs while in the other approach, the hierarchy is enforced by the initial design. Also each module is responsible for certain part of the overall behavior. However we can not assert that each ADF in the first approach corresponds to a simple behavior. However, it can be interesting idea to study the existence of relation between some of the automatically defined functions, especially those which have acquired high fitness or used by many individuals, and some of the simple tasks performed by the robot. This can be achieved by observing the robot behavior during the execution of segments of the program tree that call these ADFs.

- **The GP design.** The first approach uses a single GP process to evolve both the main program and the its functions, while in the second approach we have a number of evolution processes equal to the number of the behavior modules plus the number or arbitrator modules (whether it uses action selection or other behavior coordination methods).
  The fitness function in the first approach usually reflects the overall desired goals of the controller. This can be seen in equations (8.1) and (8.2). On the other hand, fitness functions in the other approach take some effort to be designed on the low level suitable for simple behavior required by each module.

Also having a special GP process for each module makes it possible to specify different terminal and function sets suitable for each module. Terminal set will contain only the sensor inputs needed by this module which can be less than all the available number of sensory inputs, thus allowing the search in a smaller space.

- **Reusability.** In the second approach, we have a different evolved program for each module thus allowing the reusability of these modules. On the other hand, the automatically defined functions of the first approach are branches within the main program tree, which are less suitable for reusing compared with the second approach.

- **Using both approaches jointly.** In this case, the human designer performs logical decomposing of the complex robot behavior and suggests a hierarchy for the controller. The modules of this architecture are allowed to be implemented by programs that use subroutines and functions. The GP will be used to explore this lower level hierarchy search for best fit program with its subroutines.

8.5 Summary

In this chapter, two possible approaches for evolving complex behaviors were discussed. In the first approach, the GP is used to explore possible hierarchy in the solution through implementing ADF and maintaining a subroutine library or using neural networks as primitive functions.

In the second approach, human programmer set the architecture of the robot controller and then the GP is used to evolve each module of this architecture. Two examples of architectures were discussed, the subsumption architecture and action selection architecture. Two experiments were presented to demonstrate this approach. The first used subsumption architecture to control a team of two robots with different capabilities to implement a cooperative behavior. The second experiment used action selection architecture to allow switching between the simpler behaviors that constitute the main behavior.

Acknowledgement

The author would like to thank S. Mary, S. Mina, S. Kirolos, S. George, and S. Mercorious for their support.

References

1. K-Team (1999) Khepera User Manual," Lasuanne, Switzerland
2. Mondada F, Franz F, Paolo I (1993) Mobile Robot Miniaturisation: A Tool for Investigation in Control Algorithm. Proceedings of the Third International Symposium on Experimental Robotics, Kyoto, Japan
3. Koza J (1992) Hierarchical Automatic Function Definition in Genetic Programming. Proceedings of Workshop on the Foundations of Genetic Algorithms and Classifier Systems 297–318, Morgan Kaufmann Publishers Inc., Vail Colorado, USA
4. Koza J (1993) Simultaneous Discovery of Detectors and a Way of Using the Detectors via Genetic Programming. IEEE International Conference on Neural Networks 3:1794–1801, San Francisco, NJ, USA
5. Koza J (1993) Simultaneous Discovery of Reusable Detectors and Subroutines Using Genetic Programming. Proceedings of the 5th International Conference on Genetic Algorithms ICGA:295–302, Morgan Kaufmann Publishers Inc.
6. Hondo N, Iba H, Kakazu Y (1996) Sharing and Refinement for Reusable Subroutines of Genetic Programming. Proceedings of IEEE International Conference on Evolutionary Computation ICEC: 565–570
7. Yanai K, Iba H (2001) Multi-agent Robot Learning by Means of Genetic Programming: Solving an Escape Problem. Proceedings of the 4th International Conference on Evolvable Systems: From Biology to Hardware: 192–203, Springer-Verlag
8. Iba H, Terao M (2000) Controlling effecective Introns for Multi-Agent Learning by Genetic Programming. Proceedings of Genetic and Evolutionary Conference GECCO: 419–426
9. Floreano D, Mondada F (1996) Evolution of Homing Navigation in a Real Mobile Robot. IEEE Transactions on Systems, Man, and Cybernetics (B) 2:396–407
10. Hulse M, Lara B, Pasemann F, Steinmetz U (2001) Evolving Neural Behaviour Control for Autonomous Robots. Max-Planck Institute for Mathematics in the Sciences, Leipzig, Germany
11. Haykin S (1999), Neural Networks: a Comprehensive Foundation, Second edition, Prentice Hall, Upper Saddle River, NJ
12. Lee K, Byoung-Tak Z (2000) Learning Robot Behaviors by Evolving Genetic Programs. 26th Annual Confjerence of the IEEE Industrial Electronics Society, IECON 2000 4:2867–2872
13. Brooks R (1986) A Robust Layered Control System For Mobile Robots. IEEE Robotics and Automation 2:14–23
14. Arkin R (1998) Behavior-Based Robotics. The MIT press, Cambridge, USA
15. Koza J (1993) Evolution of Subsumption Using Genetic Programming. In: F. Varela and P. Bourgine (eds) Toward a Practice of Autonomous Systems 110–119, Cambridge, MIT Press, USA
16. Kamio S, Hongwei L, Mitsuhasi H, Iba H (2003) Researches on Ingeniously Behaving Agents. Proceedings of NASA/DoD Conference on Evolvable Hardware 208–217
17. Maes P (1989) The Dynamics of Action Selection. Proceedings of the Eleventh International Joint Conference on Articial Intelligence IJCAI89 2:991–998
18. Lee W, Hallam J, Lund H (1997) Learning Complex Robot Behaviours by Evolutionary Approaches. 6th European Workshop on Learning Robots EWLR6 42–51

9

Automatic Synthesis of Microcontroller Assembly Code Through Linear Genetic Programming

Douglas Mota Dias[1], Marco Aurélio C. Pacheco[1] and José F. M. Amaral[2]

[1] ICA — Applied Computational Intelligence Lab
Electrical Engineering Department
Pontifícia Universidade Católica do Rio de Janeiro
R. Marquês de São Vicente 225, Gávea, Rio de Janeiro, CEP 22453-900, RJ
Brazil
(douglasm,marco)@ele.puc-rio.br

[2] Department of Electronics Engineering
UERJ — Rio de Janeiro State University
R. São Francisco Xavier, 524, Maracanã,
Rio de Janeiro, CEP 20550-013, RJ
Brazil
franco@uerj.br

This chapter considers the application of linear genetic programming in the automatic synthesis of microcontroller assembly language programs that implement strategies for time-optimal or sub-optimal control of the system to be controlled, based on mathematical modeling through dynamic equations. One of the difficulties presented by the conventional design of optimal control systems lies in the fact that solutions to problems of this type normally involve a highly non-linear function of the system's state variables. As a result, it is often not possible to find an exact mathematical solution. As for the implementation of the controller, there arises the difficulty of programming the microcontroller manually in order to execute the desired control. The research that has been done in the area of automatic synthesis of assembly language programs for microcontrollers through genetic programming is surveyed in this chapter and a novel methodology in which assembly language programs are automatically synthesized, based on mathematical modeling through dynamic plant equations, is introduced. The methodology is evaluated in two case studies: the cart-centering problem and the inverted pendulum problem. The control performance of the synthesized programs is compared with that of the systems obtained by means of a tree-based genetic programming method. The synthesized programs proved to perform at least as well, but they had

D. M. Dias et al.: *Automatic Synthesis of Microcontroller Assembly Code Through Linear Genetic Programming*, Studies in Computational Intelligence (SCI) **13**, 193–227 (2006)
www.springerlink.com

the additional advantage of supplying the solution already in the final format of the implementation platform selected: a microcontroller.

9.1 Introduction

Automatic control has played a vital role in the advancement of engineering and science. In addition to being a major asset in the area of space vehicles, missile guiding systems, robotic and other similar systems, automatic control has become an important and integral part of modern industrial and manufacturing processes. For example, automatic control is an essential element in the numerical control machine tools used in the manufacturing industries, in the design of automatic pilot systems used in the aerospace industry and in the automobile and truck projects of the automobile industry. It is also essential in industrial operations, as in the case of pressure, temperature, humidity, viscosity and flow control in the process industries. The theoretical and practical advancement of automatic control provides means by which optimal performances from dynamic systems may be obtained, by which productivity may be improved, by which the tiresome job of carrying out many repetitive routine operations manually may become less burdensome, and much more [21].

The types of difficulties encountered in the course of developing a control system may be divided according to two stages: design and implementation.

At the designing stage, the problem arises during the mathematical development of a solution to the control problem. Although many physical relations are often represented by linear equations, in most cases the relations are not exactly linear. In fact, a meticulous study regarding physical systems reveals that even the so-called "linear systems" are really linear only within limited operating ranges. In practice, many electromechanical, hydraulic, pneumatic, etc. systems involve nonlinear relations between the variables. For example, the output of a component may saturate for large input signals. Non-linearities of the quadratic rule type may occur in some of the components. Some important control systems are nonlinear regardless of the input signal values. For instance, in "on-off" control systems, the control action is either "on" or "off" and there is no linear relation between the controller's input and its output [21].

Analytical methods for solving nonlinear systems that are higher than the fourth order do not exist. Due to this mathematical difficulty, it normally becomes necessary to introduce "equivalent" linear systems as substitutes for the nonlinear ones. These equivalent linear systems are valid only within a limited operating range. Only after a nonlinear system has been approximated by a linear mathematical model can analytical tools be applied for analysis and design purposes [21].

In practice, proportional integral derivative PID controllers are used. PID controllers are useful on account of their general applicability to most control systems. In the field of continuous process control systems, it is known for a fact that PID and modified PID control structures have proved to be useful by providing satisfactory control. However, these controllers are unable to provide optimal control in a number of specific situations [21].

Solutions to optimal control problems normally involve a higher-order nonlinear function of the system's state variables. As a result, it is often not possible to find an exact mathematical solution to optimal control problems.

As for the implementation stage, there arises the problem of developing a program which, when executed by a digital system, is able to perform the control strategy that has been determined in the previous stage. The fact is that using computers or microcontrollers (MCs) to implement controllers results in considerable advantages. Many of the difficulties encountered with analog implementation may be avoided. For example, there are no problems with regard to precision or component alterations. It is easier to have sophisticated calculations in the control law and to include logical and nonlinear functions.

In the so-called embedded systems, where MCs are widely used, it is somewhat difficult to program the MC by comparison to a conventional computer because the computational power of an MC's central processing unit (CPU) is much more limited on account of its small instruction set and the little memory space available.

9.2 Survey on Genetic Programming Applied to Synthesis of Assembly

A number of papers on genetic programming applied to the synthesis of assembly language programs have been presented lately. Table 9.1 shows a taxonomy that includes some of the main studies on the evolution of assembly language programs.

9.2.1 JB Language

One of the earliest approaches to evolution of computer programs similar to machine code is the JB language [9]. Cramer formulated his method as a general approach to evolve programs, but his register machine language is in many ways similar to a simple machine code language and will thus serve as a good illustration for register machine GP. He uses a string of digits as the genome. Three consecutives digits represent an instruction. The first integer in the triple determines the type of instruction. This is similar to the syntax of a machine code instruction which has specific bit fields to determine the type of instruction. There are five different instructions or statements in JB. INCREMENT adds one to a specified register. The register number is given by the second digit of the triple. ZERO clears a register while SET assigns a value to a register. There are also BLOCK and LOOP instructions that group instructions together and enable while-do loops. JB employs a variable-length

Table 9.1. Taxonomy over the evolution of assembly language programs

System	Author	Description	Refs.
JB Language	N.L.Cramer, Texas Instruments	• Variable-size linear chromosome earlier, in tree later • One of the earliest approaches	[9]
VRM-M	L.Huelsbergen, Bell Labs.	• Fixed-size linear chromosome • Objective was to evolve loops.	[12]
AIMGP	P.Nordin, University of Dortmund	• Fixed-size linear chromosome earlier, variable size, later • Ran only on SUN SPARC platform earlier, also on Intel x86, later	[18][19]
GEMS	R.L.Crepeau, NCCOSC RDTE	• Fixed-size linear chromosome • Evolution of a "Hello world" program for the Z-80 CPU	[10]

string crossover. In part to avoid infinite loops, Cramer later abandoned this method in favor of an approach using a tree-based genome [5].

9.2.2 VRM-M

Huelsbergen has used a formal approach to evolve code for a virtual register machine (VRM) [12]. His goal was to evolve an iterative multiplication program from lower level instructions such as JUMP, COMPARE, ADD, DECREMENT and others. Execution loops are formed spontaneously from these primitives given fitness cases and a maximum number of allowed execution steps. The evolutionary algorithm also uses a linear representation of the programs (string of instructions) and a linear crossover operator exchanging substrings. His system is implemented in the functional formal language Standard Meta Language (SML). GP results are compared favorably with results using random search.

9.2.3 AIMGP

Nordin has used the lowest level binary machine code as the programs in the population [18][19]. Each individual is a piece of machine code that is called and manipulated by the genetic operators. There is no intermediate language or interpreting part of the program. The machine code program segments are invoked with a standard C function call. The system performs repeated type cast between pointers to arrays for individual manipulation and pointers to functions for the execution of the programs and evaluation of individuals' fitness. Legal and valid C-functions are put together, at run time, directly in memory by the genetic algorithm. This system was formally known as

CGPS (Compiling Genetic Programming System), but later it was renamed to AIMGP (Automatic Induction of Machine code with Genetic Programming). Figure 9.1 illustrates the structure of an individual.

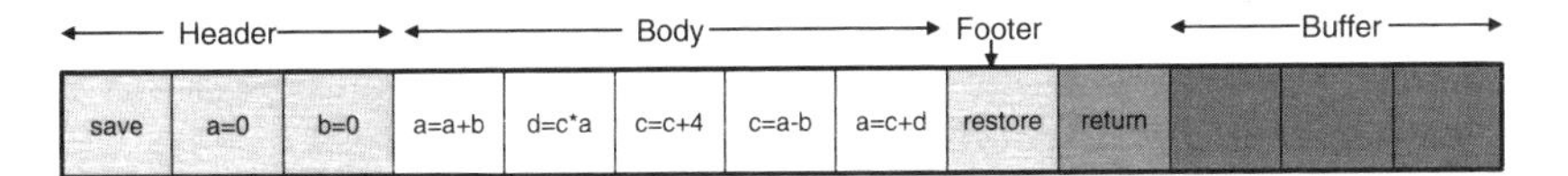

Fig. 9.1. Structure of a program individual in AIMGP

The header deals with administration necessary when a function gets entered during execution. It is often constant and can be added at a initialization of each individual's machine code sequence. The footer cleans up after a function call. The return instruction follows the footer and forces the system to leave the function and to return program control to the calling procedure. The function body consists of the actual program representing an individual. A buffer is reserved at the end of each individual to allow for length variations. The genetic operators must be prevented from changing the header, the footer and the return instruction.

9.2.4 GEMS

One of the most extensive systems for evolution of machine code is the GEMS system [10]. The system includes an almost complete interpreter for the Z-80 8-bit microprocessor. The Z-80 has 691 different instructions, and GEMS implements 660 instructions, excluding only special instructions for interrupt handling and so on. It has so far been used to evolve a "hello world" program consisting of 58 instructions. Each instruction is viewed as atomic and indivisible; hence crossover points always fall between the instructions in a linear string representation [5]. Figure 11.2 illustrates the crossover and mutation process used in GEMS, where a new offspring is created by exchanging a block of instructions between two fit parents and by mutating its result.

9.2.5 Discussion

The AIMGP system was developed mainly with a view to the achievement of a good velocity performance. This performance is made possible on account of the fact that the individual is evaluated through its direct execution by the CPU of the evolutionary platform, thus eliminating any type of intermediate stage, such as, for example, the compilation or the interpretation stages. The disadvantage of this approach is that the system becomes less flexible since the programs can only be evolved to the platform on which the evolutionary algorithm is running.

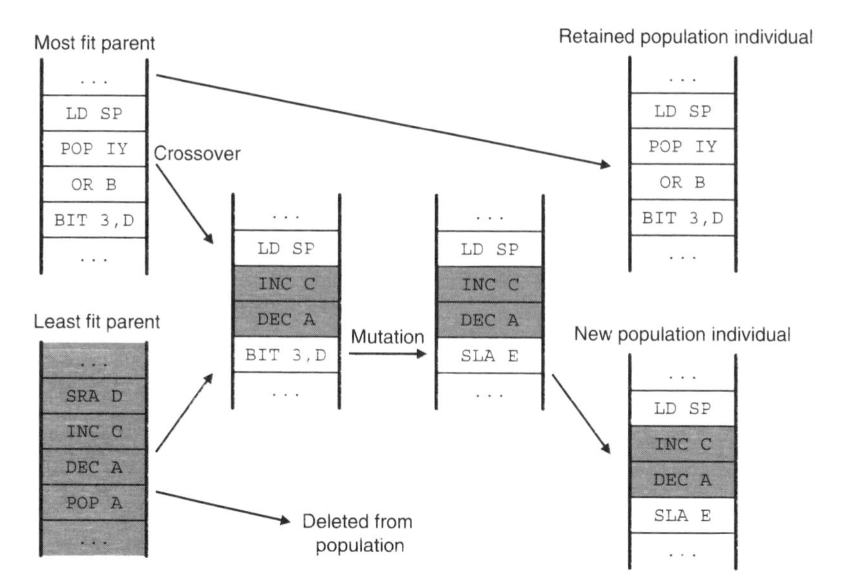

Fig. 9.2. The GEMS crossover and mutation process

The remaining systems mentioned above, however, did not aim to excel in evolutionary velocity, but rather in offering the possibility of evolving programs to a specific type of assembly language. The system presented in this chapter also belongs to this category because its purpose was the creation of programs for an existing PIC microcontroller. This approach is somewhat similar to that of the GEMS system since the GEMS also evolves programs for an existing device, which, in its case, is a Z80 CPU.

In the same manner as the AIMGP system, the system that has been developed in this paper made use of a variable-size linear chromosome. This makes it possible to try to optimize the code of the evolved programs.

9.3 Design with Microcontrollers

9.3.1 Microcontrollers

As integrated circuits (ICs) became cheaper and microprocessors (MPs) began to emerge, the simpler MPs began to be used in the implementation of dedicated tasks, such as printer control, velocity adjustment, step motor control, elevator controls, etc.

However, implementing a control system involved assembling very large circuits and this often represented a price increase. The cost of such applications depended on the price of the MP and of the peripherals (ROM and RAM memories, ports, A/D and D/A converters, etc.), as well as on the number of connections and the circuit board size.

For the purpose of reducing the price, the idea of placing these peripherals and the CPU within a single IC was conceived. This would not only reduce the price and the size of the printed circuit, but would also increase the system's credibility. Moreover, it is not necessary for a control-dedicated CPU to be very fast, nor does it need to have such an extensive and powerful instruction set. Instructions are not necessary in work involving floating points, strings, or vectors. The addressing mechanisms must also be simple. In other words, it is possible to simplify the CPU. And that is how microcontrollers came on the scene - they are simple, cheap and efficient.

9.3.2 Time-Optimal Control

Control-related problem-solving is a task that involves conducting the state of a system from a given initial state to a specific final state. The state of the system is described by a set of state variables. The transition from one state to a next one is obtained by means of a control variables set. Thus, the control problem consists of selecting the values of these variables so as to cause the system to move towards the specified final state. The goal of an optimal control is to perform control at an optimal cost, where cost can be measured in terms of time, fuel, energy or some other type of measurement. What is important in a minimum-time problem class is to transfer the state of a system to a specific final state as quickly as possible [4].

Solutions to control and optimal control problems normally involve a higher-order nonlinear function of the system's state variables. As a consequence, many times it is not possible to find an exact mathematical solution to control problems. Moreover, in the case of practical problems, it often happens that the mathematical model of the system itself is not available and it may become necessary to obtain the control based on a set of data that have been obtained empirically.

It is also important to point out that although many of the physical relations in the system that is to be controlled are often represented by linear equations, in most cases the real relations are not exactly linear [21].

As already said, the use of computers for the implementation of controllers presents considerable advantages [3]. There are, however, certain types of systems in which it is not possible to make use of computers. They are the so-called embedded systems, where MCs are widely employed. MCs also present the advantage of providing computational resources for the control system that is to be implemented, though in a much more limited manner. It can be said that genetic programming has much to contribute in this scenario. If on the one hand, it represents a means by which to automatically synthesize programs that implement control and optimal control strategies, even in the case of problems where it is not possible to obtain a mathematical solution, on the other, it is able to do so specifically for a platform that has limited computational resources, as in the case of the MC.

9.4 Microcontroller Platform

This section contains a brief presentation of the PIC18F452 Microchip [16], an 8-bit RISC microcontroller, with a simple architecture, largely employed and that has been adopted as a platform in the automatic synthesis of assembly code.

The architecture of this MC is of the Harvard type, in other words, there are two internal buses, one containing data, the other, instructions. The program and the data therefore occupy different memories. The program memory is of the flash type with storage capacity of 32 Kbytes, that is, 16,384 16-bit instructions.

The data RAM is responsible for the storage of the file registers and of the special registers. The area of this memory, which is related to the general-purpose registers, stores the variables defined by the programmer which are to be written and read by the program. In turn, the area related to the registers for special use can be read and written by both the program and the hardware since the MC makes use of these registers for executing the program and for processing the Arithmetic-Logic Unit (ALU). These are the registers that contain, for example, the status flags. This memory has a storage capacity of 1,536 bytes.

The "W" accumulator is an 8-bit register that directly stores the ALU operation results and it is also addressable.

There is a hardware multiplier in the ALU (8 x 8 multiply), which brings with it the advantages of a better computational throughput and a reduction in the size of the necessary code in multiplication algorithms. Multiplication is carried out, without any signal, between two 8-bit registers and their 16-bit result is stored in two specific 8-bit registers: "PRODH" and "PRODL".

In addition, there are 5 input/output ports (I/O ports), where port "A" has 7 bits, ports "B", "C", and "D" have 8 bits each and port "E", 3 bits. All these bits can be configured individually as input or output pins by means of the correct configurations of their respective special registers. The ports are addressable in the data bus.

This MC has a set of 75 instructions, 22 of which were selected for the experiments contained in this work and are indicated in table 9.2. Operand "f" assigns the data memory register that the instruction is to use, while operand "d" assigns the destination of the result of this operation. If "d" is zero, then the result will be stored in "W", and if it is one, in the register assigned by "f".

These operations were chosen because the control problems discussed in this paper make it necessary to basically deal with arithmetical and conditional operations. The selection of a subset of operations that is suited to the nature of the problem is indispensable as an initial stage if GP is expected to yield a good performance.

Table 9.2. Instruction subset of the PIC18F452

Mnemonic and Operands		Description
`ADDWF`	`f, d`	Add W and f
`ADDWFC`	`f, d`	Add W and the Carry bit to f
`CLRF`	`f`	Assigns value zero to f
`CPFSEQ`	`f`	Compares f with W, skips next instruction if f = W
`CPFSGT`	`f`	Compares f with W, skips next instruction if f >W
`CPFSLT`	`f`	Compares f with W, skips next instruction if f <W
`DECF`	`f, d`	Decrements f
`DECFSZ`	`f, d`	Decrements f, skips next instruction if f = 0
`INCF`	`f, d`	Increments f
`MOVF`	`f, d`	Copies the content of f
`MOVFF`	`fs, fd`	Copies the content of f_s in f_d
`MOVWF`	`f`	Copies the content of W in f
`MULWF`	`f`	Multiplies W with f
`NEGF`	`f`	Negates f (two's complement)
`RLCF`	`f, d`	Rotates f to the left through the Carry bit
`RLNCF`	`f, d`	Rotates f to the left without the Carry bit
`RRCF`	`f, d`	Rotates f to the right through the Carry bit
`RRNCF`	`f, d`	Rotates f to the right without the Carry bit
`SETF`	`f`	Assigns value one to all the bits of f
`SUBWF`	`f, d`	Subtracts W from f
`BTFSC`	`f, b`	Tests bit b of f, skips next instruction if b = 0
`BTFSS`	`f, b`	Tests bit b of f, skips next instruction if b = 1

9.5 Linear Genetic Programming

The emergence of GP in the scientific community arose with the use of the tree-based representation, in particular with the use of LISP language in the work of Koza [14]. However, GP systems manipulating linear structures exist [7], which have shown experimental performances equivalent to Tree GP.

In linear genetic programming, the programs are represented by a linear string of instructions, of variable size. When compared with traditional GP, which is based on tree structures, its main characteristic is that programs are evolved in an imperative language such as C [7] and machine code [18], and not in expressions of a functional programming language, such as LISP [14]. It is for this reason that this type of GP is the best option, since assembly language is also an imperative language.

The implementation of the linear GP proposed in [7], for example, represents an individual program as a variable-size list of instructions in C which operate with indexed or constant variables. In linear GP, all the operations, such as, for example, $a = b + 1$, implicitly include an assignment to a variable. After the program is executed, the output values are stored in the assigned variables, differently from the tree-based GP, where the assignments and multiple outputs must be explicitly incorporated through the use of an extra indexed memory and of special functions for a manipulation of this type.

9.5.1 Evolution of Assembly Language Programs

Assembly language programming is frequently used when there is a need for very efficient solutions, as is the case of applications that are strongly restricted in terms of run time and memory use. In general, the reasons for evolving assembly language as opposed to high-level languages are similar to the reasons for the use of manual programming in assembly language through the use of mnemonics or in machine code.

This is the level at which the most efficient optimization is obtained. It is the lowest level to which a program can be optimized, and it is also where it is possible to obtain the highest gains. Optimization may occur as a result of velocity, space, or both. Small individuals may be promoted by means of a penalty for length, which is commonly known as parsimony pressure [19]. Similarly, this pressure may be applied considering the run time of the instructions.

It is usually considered difficult to learn, program and master assembly language. Sometimes it may be easier to let the computer evolve small programs in assembly language than to learn to master the technique of programming at this level. This is acceptable, not only in case a high-level compiler is not available, but also when the use of assembly language is preferred for one of the two above mentioned reasons.

9.6 System for Automatic Synthesis of Microcontroller Assembly

This section describes a System for Automatic Synthesis of a Microcontroller Assembly that makes it possible to evaluate the proposed genetic approach through benchmark case studies. The system also includes a simulator of the plant to be controlled and a simulator of the microcontroller.

The main routine controls the system by continuously evaluating the generated assembly programs, which are executed using both plant and MC simulators together, and then submitted to the genetic operators in the evolutionary kernel.

Figure 9.3 shows the block diagram of the system, and the operation of each module is detailed in the subsequent subsections.

9.6.1 Evolutionary Kernel

This module is responsible for the evolution of the programs themselves and therefore implements the selection and evolutionary algorithms as well as the

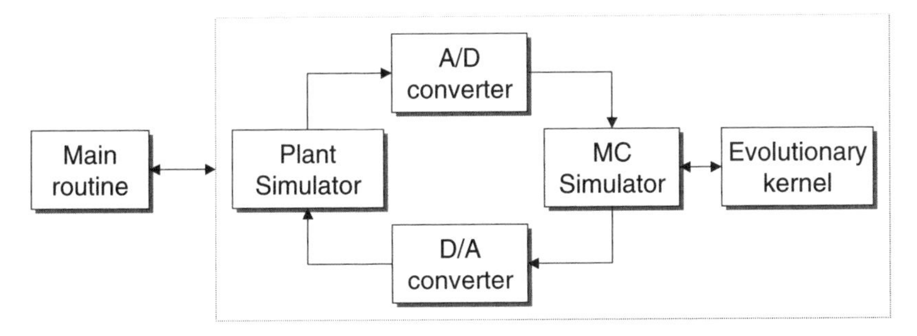

Fig. 9.3. Block diagram of the system

genetic operators. The module operates by creating the initial population, selecting individuals and applying the genetic operators to these individuals. The next subsections contain a detailed description of how this module operates.

Representation of the Individuals

Given the fact that the objective of this system is to evolve programs in assembly language, it was decided that these programs would be represented by linear chromosomes [19], where each gene represents an instruction that is composed of the operation and of one or two operands.

In practice, this chromosome was implemented by a vector, whose length was equivalent to the length of the program, with an instruction format structure.

The individuals are created at the stage in which the population is initialized, which is the first stage of the GP, and makes use of a process that is similar to the one employed in [19]. First, an individual is created through the random selection of a length between the minimum initial length and the maximum initial length parameters. Next, all its genes are filled in with randomly generated instructions based on the syntax defined for assembly language. This process is repeated for all the individuals of the population.

Crossover

Linear crossover consists of an exchange of code segments between two parents, according to the graphic demonstration in Figure 9.4.

A random string of instructions is selected in each parent. In Figure 11.4, the selected instructions are highlighted in gray. These sequences are exchanged between the two parents and as a result, the descendent individuals are generated.

For the length of these individuals, the heuristic used in [6], which demonstrates by means of experiments that unrestricted exchanging of subprograms

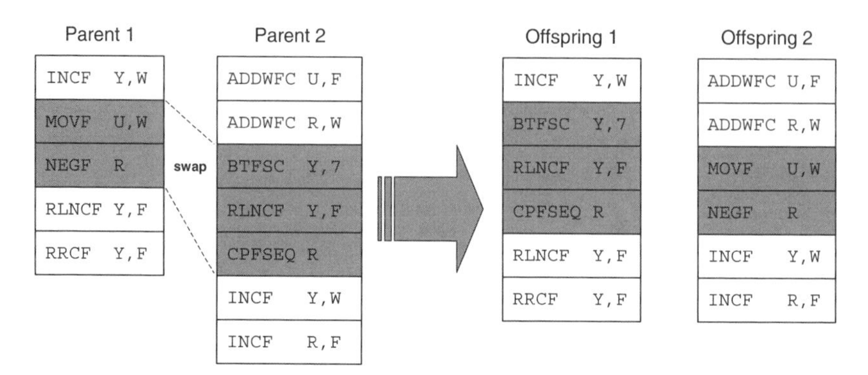

Fig. 9.4. Crossover of two individuals

is not advisable, was adopted. The fact is that when large program segments are exchanged, the crossover becomes too destructive in evolutionary terms. The approach that was suggested and adopted consisted of limiting the maximum size of these segments and after this work was analyzed, a maximum value of 4 instructions was adopted.

Mutation

Mutation acts upon only one individual at a time. After crossover occurs, each offspring is submitted to the mutation that has a lower level of probability than the crossover. The crossover and mutation probabilities are the execution parameters of a GP.

When an individual is selected for mutation, the mutation operator initially selects one of the individual's instructions and changes its content. The type of change is selected in a random fashion and may consist of changing the operation or one of the two operands. This approach was also obtained from [6], where it is called "micromutation" and its purpose is to smoothen the effect of this operator during evolution. Figure 9.5 illustrates how this operator works and shows the parent with examples of two possible offspring: the one on the left would be the result of a random choice in favor of changing the operation and the one on the right, of changing the first operand.

An alternative approach to this mutation for linear GP may be the so-called "macromutation" [6], where a randomly selected gene is simply removed or a new randomly created gene is inserted in a random position. Another possible approach consists of exchanging a code segment for an entirely new, randomly selected one [5]. But the experiment results obtained in the linear GP-related literature demonstrate that micromutation is more efficient because it is not as destructive as the others.

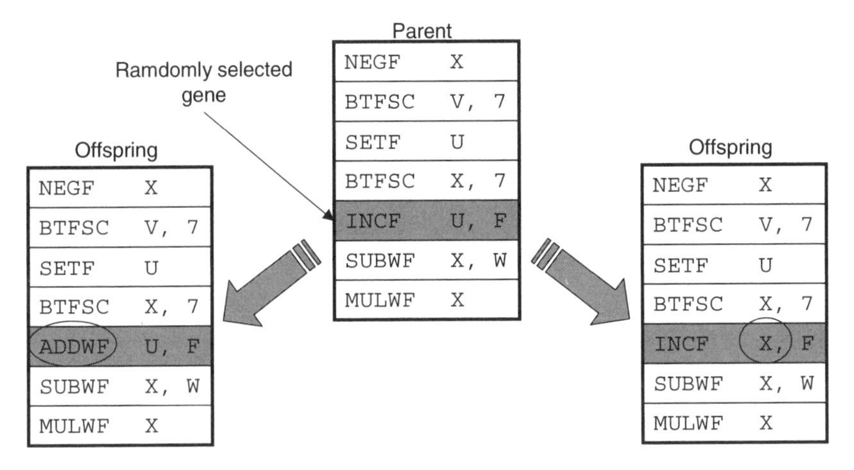

Fig. 9.5. Mutation on an individual

Selection

This system has adopted the tournament selection genetic algorithm [19]. Tournament selection is not based on a competition among all the individuals of a generation, but rather among the individuals of a subset of the population. In this type of tournament, the number of individuals, which defines the size of the tournament, is selected at random before a selective competition begins. The objective in this case is to allow the characteristics of the best individuals to replace those of the worst individuals. In the smallest tournament possible, two individuals compete with each other and the best one of these two is reproduced and may undergo mutation. The result of the breeding is returned to the population and replaces the tournament loser.

The size of the tournament makes it possible for researchers to adjust selection pressure. Small tournaments cause low selection pressure, while large ones cause high pressure. Based on the system developed in [19], a tournament size equal to four was selected.

Tournament selection has become quite popular as a selection method mainly because it does not require centralized fitness comparisons among all the individuals. This not only significantly accelerates evolution, but also provides a simple way of making the algorithm become parallel. With fitness-proportional selection, the communication overhead between evaluations may be even higher [5].

Evolution

The evolutionary algorithm selected was the so-called steady-state algorithm, which is based on the tournament selection model. It should be pointed out that the steady-state algorithm used in GP is different from the one used in genetic algorithms. In this approach, there are no fixed intervals between generations. Instead, there is a continuous flow of individuals that meet, mate and breed. The offspring replace individuals that exist in the same population. This method is easy to implement and presents certain advantages with respect to efficiency and to possible expansion through parallelism. Good results for convergence in general have been reported and the method is becoming increasingly popular within the research community. The evolutionary algorithm that was specifically developed for the evolutionary kernel of the system discussed in this paper is described below:

1. Four individuals from the population are randomly selected.
2. The four individuals are grouped in two pairs. For each pair, the two individuals are evaluated and their resulting fitness are compared with one another.
3. The two winners, one from each pair, are probabilistically submitted to mutation and crossover.
4. The two losers, one from each pair, are replaced by the offsprings of the two winners.
5. Steps 1 to 4 are repeated until the maximum number of cycles is achieved.

This is called the steady state approach because the genetic operators are applied asynchronously and there is no centralized mechanism for explicit generations. However, steady state GP results are usually presented in terms of generations. Actually, the steady state generations are the intervals that occur during evolution which, one might say, correspond to the generations in a generation-based GP. These intervals are computed each time that fitness is evaluated by the same number of individuals that the generation-based GP population would have. [13] contains experiments and detailed references with respect to generation-based GP versus steady state GP.

9.6.2 Plant Simulator

Two simulators were developed, one for each case study. Each simulator was implemented by means of a function that takes the state variables, through reference, and the value of the control variable. At each specific time interval, the values of the plant's state variables are changed as a result of the control variable. The plant's control variable, in turn, represents the control action itself and results from the execution, by the MC simulator, of the program that is being evaluated.

The kernel of this function contains the dynamic equations that mathematically model the physical system that is being simulated.

9.6.3 Microcontroller Simulator

This simulator was implemented by means of a function that takes the values of the plant's state variables and returns the new value of the control variable.

That is, at the end of each plant simulation time interval, the program that is being evaluated receives the updated state variable values originating from the plant simulator, is executed from beginning to end, and the value of the variable that performs the control action at that given instant is returned.

As has been mentioned in section 9.3, out of the 75 instructions contained in the PIC18F452, only those pertaining to the scope of the case studies were implemented and the result was a total of 22 instructions.

9.6.4 A/D and D/A Converters

These converters are responsible for establishing an interface between the analog magnitude values with which the plant simulator works and their digital equivalents used by the MC simulator.

A/D converters are functions that convert the analog values of the plant's state variables, originating from the plant simulator, to their respective digital values, which are supplied to the MC simulator. In other words, these modules simulate the operation of physical A/D converters, which convert the analog magnitude values obtained by their respective sensors to their equivalent digital values, which are furnished to the MC.

In turn, the D/A converter is a function that converts the digital value of the control variable, originating from the MC simulator, to its respective analog value, which is supplied to the plant simulator. Hence, this module simulates the operation of the physical D/A converter, which converts the digital value of the MC's output port to its respective analog value, which controls the action of the plant's physical control actuator.

9.6.5 Overview

The evolutionary process begins with the creation of an initial population, its basis being the syntax defined for assembly language. Next, the evolutionary kernel selects the individuals that are to be evaluated, evaluates them and submits the best ones to the genetic operators. Figure 9.6 shows how the system operates.

When the evaluation of an individual begins, the plant simulator is placed in a given initial training state. The individual is then repeatedly executed, once for each plant simulation time interval, and this process is repeated until the last time interval for this simulation is achieved. In turn, the execution of an individual is carried out by the PIC simulator, which receives the program and the current values of the input and state variables, and returns the value of the control variable for that given plant simulation time interval. This control variable is the element that actually acts on the plant by performing its control. This process is repeated for each initial training state until the last one is reached. In this manner, an individual's fitness level, which measures how well it performed the control of that plant, is obtained.

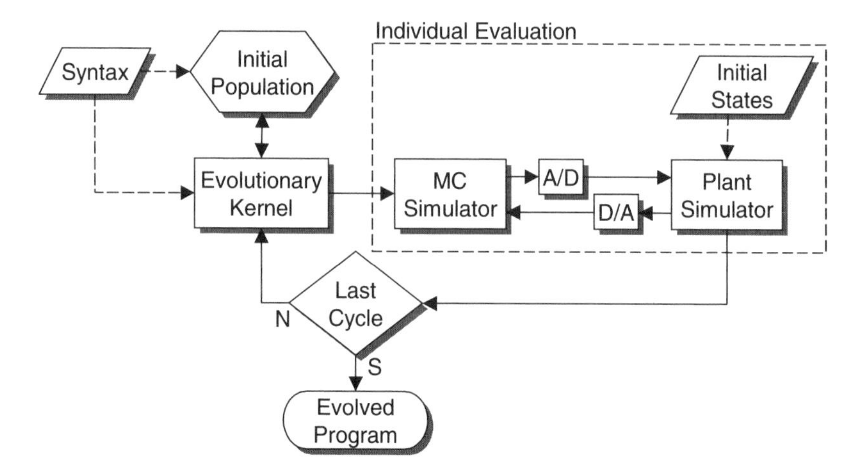

Fig. 9.6. Operation of the synthesis system

This evolutionary process is repeated cycle after cycle until the last one is reached, which is when a specific program resulting from that experiment is determined, and this program is the individual that presented the best fitness obtained among all the cycles that have been executed up to then.

9.7 Case Studies

Two benchmark case studies have been selected: the cart-centering problem, which makes it possible to check if the time-optimal control strategy has been obtained, because this strategy is known analytically; and the inverted pendulum problem, a system that is difficult to control because it is inherently unstable and has three state variables. The case studies were evaluated in relation to another approach found in the literature. All the experiments were run on a PC computer with a Pentium IV 2.8 GHz processor.

9.7.1 Cart Centering

The computational intelligence area is quite familiar with the cart-centering problem. Koza [14] successfully applied GP to demonstrate that GP was capable of providing a controller that would be able to center the car in the shortest amount of time possible.

This problem is also known as the double integrator problem and appears in textbooks that introduce the concept of optimal control as the application of the Pontryagin principle, as in [4]. Considerable research on the theory of this problem has been carried out and it is possible to calculate the best theoretical performance although the development of an expression that produces this performance is still a nontrivial activity.

Thus, the cart-centering problem involves a car that can move towards the left or towards the right on a frictionless one-dimensional track. The problem consists of centering the car, in minimum time, through the application of a fixed magnitude force (a bang-bang force) in such a way as to accelerate the car towards the left or towards the right. In Figure 9.7, the car's current position $x(t)$ at instant t is negative and its velocity $v(t)$ is positive. In other words, the position $x(t)$ of the car is to the left of its origin (0.0) and the current velocity $v(t)$ is in the positive direction (towards the right). The force F is positive, that is, the bang-bang force is being applied to the car by the motor in such a way as to accelerate it positively (towards the right).

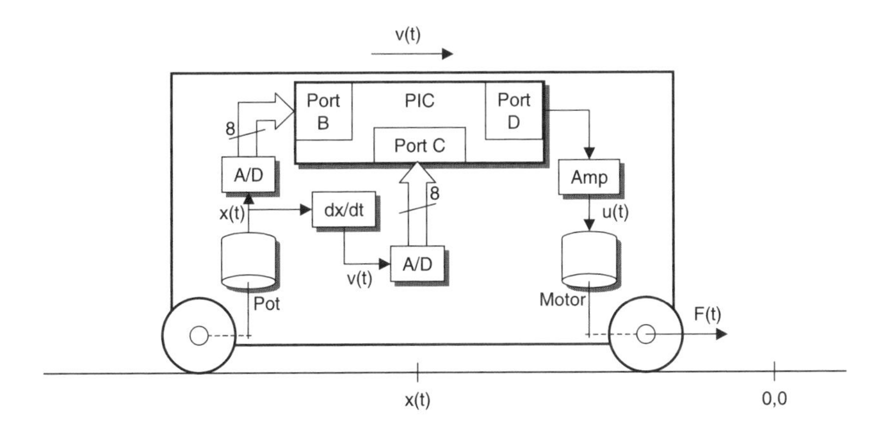

Fig. 9.7. The cart-centering problem controlled by a PIC

In this hypothetical experiment, the PIC takes car's velocity and position as its inputs and returns the control action according to the values of these inputs. The value relative to position $x(t)$ is obtained by means of a potentiometer (attached to one of the wheels) that is fed and calibrated so as to provide a value between -0.75 and 0.75 volts to represent the value of $x(t)$ between -0.75 and 0.75 meters. This analog voltage value is then converted to digital and is provided to the PIC through port B. For obtaining the value relative to the velocity $v(t)$ of the car, the signal originating from the potentiometer passes through a calibrated differentiator so as to provide a value between -0.75 and 0.75 to represent the value of $v(t)$ between -0.75 and 0.75 meters per second. This analog signal is then converted to digital and is provided to the PIC through port C. The control $u(t)$ of the car is performed by the PIC through the pin relative to the least significant bit of port D. Its value is provided to the current amplifier which, upon receiving the low logical value (0V), activates the motor so that a force in the negative direction is imparted to the car, and upon receiving the high logical value (5V), activates it imparting a force in the positive direction to the car.

The cart-centering problem is an optimal control problem. Problems of this type involve a system that is described by state variables. The control variable's action causes a change in the system's state. The goal is to select a control variable value that will get the system to go to a specified final state with an optimal cost. In the problem at hand, the state variables are $x(t)$ and $v(t)$, while the control variable $u(t)$ represents the direction in which the force will be applied. The cost to be minimized is time, that is to say, this is specifically a time-optimal problem. The final state to be achieved is the state in which the car is at rest ($v(t) = 0$) and centered at its origin ($x(t) = 0$). Hence, the objective of this problem is to choose a sequence of values for $u(t)$ so as to conduct the system to the final state in a minimum amount of time.

The bang-bang force is applied at each time interval and causes a change in the system's state variables. When this force $F(t)$ is applied to the car at instant t, the car accelerates in accordance with Newton's law in the following manner:

$$a(t) = \frac{F(t)}{m} \tag{9.1}$$

where m is the car's mass. Then, as a result of this acceleration $a(t)$, the velocity $v(t+1)$ of the car at instant $t+1$ (which occurs a short time τ after the moment t) is

$$v(t+1) = v(t) + \tau a(t) \tag{9.2}$$

where τ is the sampling interval.

At instant $t+1$, the position $x(t+1)$ of the car becomes

$$x(t+1) = x(t) + \tau v(t) \tag{9.3}$$

Hence, the change that occurs in the value of the control variable at instant t causes the system state variables to change at instant $t+1$. The value of the control variable $u(t)$ multiplies the value $|F|$ of the force F by +1 or by -1.

The time-optimal solution obtained in [4] is: for whatever current position $x(t)$ and whatever current velocity $v(t)$, apply a bang-bang $F(t)$ force in order to accelerate the car in the positive direction if

$$-x(t) > \frac{v(t)^2 Sign(v(t))}{\frac{2|F|}{m}} \tag{9.4}$$

or, on the contrary, apply a bang-bang force F to accelerate the car in the negative direction. The sign function returns +1 for a positive argument and -1 for a negative argument. Based on the values used in [14], the car's mass m has a value of 2.0 kilograms and the force F a value of 1.0 Newton.

Evaluation Function

An individual's fitness is the total amount of time needed for centering the car, given some initial condition points (x, v) in the domain that has been specified for the problem. For the purpose of computing this total time, the individual must be submitted to a test in which it must be executed for all the training cases. The training cases are evenly spaced selected values of the state variables within the specified domain. In this case, $k = 25$ pairs of values were selected, 5 for $x(0)$ and 5 for $v(0)$, and both variables were given values: -0.750, -0.375, 0, 0.375 and 0.750. In this manner, an individual's performance is tested for each one of these 25 training cases and the total time is calculated. Figure 9.8 contains the flowchart of the evaluation function for a given individual of the population.

First, the function initializes the individual's fitness at zero. Next, the outermost loop, which is responsible for sweeping all the 25 training cases, is initialized. Once within the loop, the values of variables x and y are converted from analog to digital so that they may later be supplied to the individual that is being evaluated. The value of these two variables lies between -0.75 and 0.75 and they may take values between 0 and 225 in the digital domain. The two's complement representation was used for the negative values of the analog magnitudes by way of taking advantage of the fact that most PIC instructions operate in this manner. Thus, the digital value returned by the A/D function according to the analog value is

$$value_{digital} = \begin{cases} int\left(\frac{127}{0.75} value_{analog}\right) & \text{if } value_{analog} \geq 0 \\ \sim \left[int\left(\frac{127}{0.75} \left|value_{analog}\right|\right)\right] + 1 & \text{if } value_{analog} < 0 \end{cases} \tag{9.5}$$

that is, if the analog value is less than zero, the digital value is obtained from the module of its value, which is multiplied by the proportional fraction. The bits of the integer part of this last operation are inverted one by one and the resulting value of this operation is added to one, and in this manner, the procedure for obtaining the two's complement is completed.

The innermost loop is then initialized. This loop is responsible for simulating the plant that is being controlled by the PIC, for a maximum period of 10 seconds for each training case. If the 10-second limit is reached, it means that the individual being evaluated represents a control program that did not succeed in centering the car in less than 10 seconds and that it therefore committed a timeout for the current training case. The number of timeouts can therefore achieve a maximum value of 25 for each individual and this value is also stored in its structure.

This simulation occurs at intervals of 0.02 second, that is, this loop is executed up to a maximum of 500 times. At each one of these intervals, the state variables are updated according to their previous values and to the value of the control action performed by the PIC. In other words, the PIC simulator

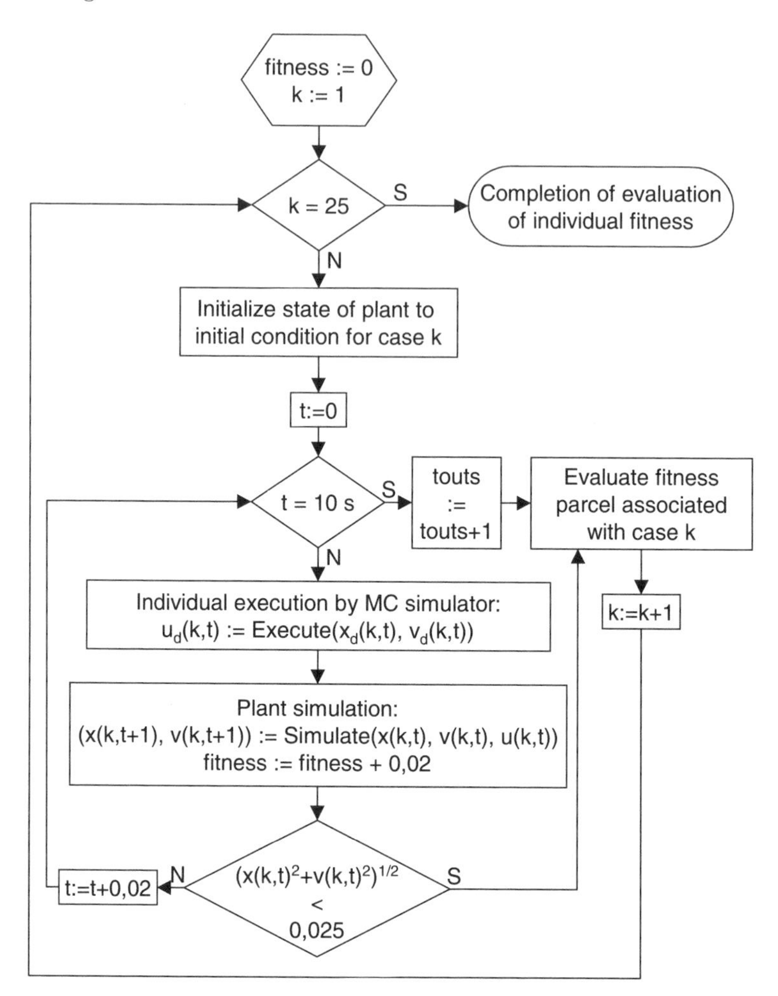

Fig. 9.8. Flowchart of the evaluation function for the cart-centering problem

takes the digital values $x_d(k,t)$ and $v_d(k,t)$ of the state variables as its inputs and returns the value of the control variable $u_d(k,t)$ as its output. The value of the control action $u(k,t)$ that will be supplied to the plant simulator is obtained in the following manner: if the value of $u_d(k,t)$ is 1, then the value of $u(k,t)$ will also be 1 because the force to be applied is 1 Newton in the positive direction; if the value of $u_d(k,t)$ is 0, then the value of $u(k,t)$ will be -1 because the force of 1 Newton will be applied in the negative direction. The plant simulator, in turn, takes the state variables $x(k,t)$ and $v(k,t)$ and the control variable $u(k,t)$, returning the updated values $x(k,t+1)$ and $v(k,t+1)$ of the state variables, which are obtained through equations 9.1, 9.2 and 9.3.

Upon completion of one more time interval, the value of 0.02 second is added to the fitness, which will have a maximum final value of 10 seconds.

As a last processing procedure for this innermost loop, a test is performed for the purpose of checking if the control action has placed the plant in the final state of equilibrium desired, which occurs when the square root of the sum of the squares of position and velocity is lower than 0.025. Otherwise, the simulation continues to be executed for the given training case k until the system is put in the state of equilibrium or until the maximum time limit of 10 seconds has been reached. At the end of this loop, the fitness parcel relative to the current case k is obtained. This simulation loop is executed for all the training cases and once all theses cases have been completed, the individual's fitness is finally obtained as being the sum of all theses parcels.

Experiment

Table 9.3 summarizes this experiment and describes its main features and parameters.

Table 9.3. Summary of the evolutionary experiment with the cart-centering problem

Objective:	To synthesize an assembly language program that enables a PIC18F452 MC to control a car on a frictionless track so as to center it in a final central position.
Terminal set:	X (position), V (velocity) and U (control); A1 and A2 (auxiliary); PRODH and PRODL (multiplier hardware registers); STATUS (flag register).
Function set:	The 22 PIC instructions described in Table 9.2.
Training cases:	25 initial condition points (x,v), selected and evenly spaced, between -0.75 and 0.75. 5 values for x and 5 values for v (-0.750, -0.375; 0; 0.375, 0.750).
Hits:	Number of training cases that did not cause a timeout.
Fitness:	Sum of the time, over the 25 training cases, taken to center the car. When a training case causes a timeout, the contribution is of 10 seconds.
Evaluation:	The least number of timeouts. As a first tie-breaking criterion, the lowest fitness. As a second tie-breaking criterion, the shortest length of the individual.
Main parameters:	M = 500. Steady state. Number of individuals processed: 400,000, which is equivalent to G = 800. $p_c = 0.8$, $p_m = 0.4$.

In [14], the evaluation of the individual is based solely on its fitness. This approach was also adopted in this paper at first, but upon completion of the first experiments, a disadvantage was noticed. When evolution begins, even the best individuals are not capable of centering the car for the 25 training cases, that is to say, the initial number of timeouts they present is higher than 20. In this approach, the authors start from the assumption that as fitness diminishes during evolution, the number of timeouts also tends to diminish. This was in fact verified in the first experiments that were carried out, but only as a tendency, not as a rule. In other words, despite the fact that in some parts of the evolution the number of timeouts diminished as the fitness also diminished, it was observed that many times the reduction of fitness led to increased timeouts. This revealed that the GP was evolving individuals that were less fit, even if this represented individuals with a higher number of timeouts, which indicates that an individual is less qualified to deal with the training cases. Therefore, an approach was adopted in which the individuals are evaluated with regard to the number of timeouts, where the tie-breaking criterion is fitness itself. This approach significantly improved the GP's evolutionary performance without affecting the experiment's initial objective, which is to promote the evolution of time-optimal control programs.

In this paper, a heuristic on evaluation was developed for the purpose of encouraging the evolution of individuals with optimized code sizes. This heuristic capitalizes on the tournament selection approach that has been adopted, which always compares the individuals' fitness in pairs. When it is observed that the individuals have the same fitness value, their lengths are compared and the one that has the lower value wins the tournament. In this manner, the individual that presents the same performance upon solving the problem, but that does so with a smaller code, is rewarded in detriment of the other.

The experiments revealed that a smaller population causes faster convergence, but with a higher final fitness value, while a large population converges more slowly, but reaches a lower final fitness value. This fact was also verified in [18]. In this manner, it was decided that M = 500 would be the value for population size in this experiment. It was determined, also by experimentation, that the crossover and mutation rate values that provided the best evolutionary performance were, respectively: $p_c = 0.8$ and $p_m = 0.4$.

During the analysis of the evolutionary curve of some of the executions of this experiment, it was noticed that convergence occurs before cycle number 100,000, which is equivalent to the evaluation of 400,000 individuals since 4 individuals are evaluated in each cycle.

Figure 9.9 shows a graph of a preliminary experiment in which the length of the individuals was not used as a tie-breaking criterion. Figure 9.10, in turn, shows the graph of the experiment that used the heuristic on the length of the individuals. Both graphs represent the average of the results of 10 executions. It may be noticed that in the first experiment (figure 9.9), the code size went on growing until it finally reached a near-90 value. This is a familiar phenomenon in GP called "bloat" [5], and it is also one of problems that the literature in this area discusses the most. What happens is that the average size of the individuals tends to increase rapidly during the evolutionary process, as, for example, in [14]. Upon observation of the evolutionary graph of the second experiment (figure 9.10), it may be noticed that the bloat problem was satisfactorily controlled with the use of this heuristic without hampering the system's evolutionary performance.

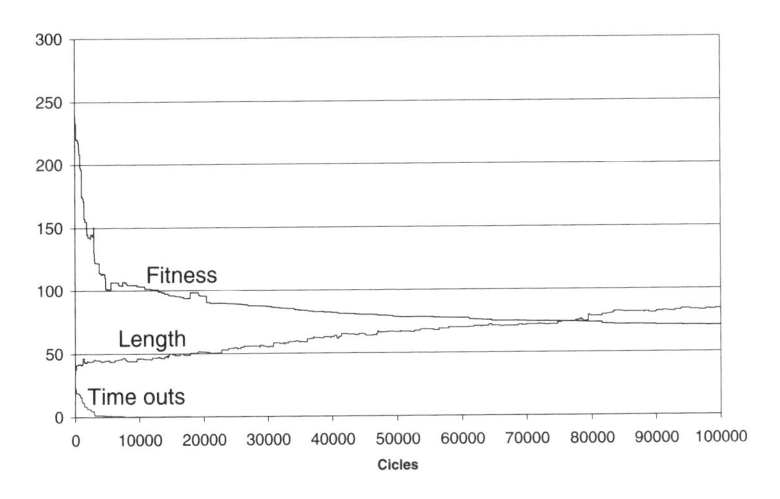

Fig. 9.9. Evolution of the cart-centering problem, without control over the length of the individuals

The best individual that was evolved in all the executions presented a fitness of about 65 seconds for the 25 training cases. In order to find out whether this individual represented the optimal solution to the problem, an electronic spreadsheet was used for the purpose of comparing the values that the program assigned to the control variable $u(t)$ with the values that it should assign according to equation 9.4, which describes time-optimal control. This comparison was carried out for all the 256 points that variables x and v would be able to assume in the digital domain, which would be equivalent to the -0.75 and 0.75 variation in the analog domain. Actually, the values of $u(t)$ obtained by the evolved program, for all these 65,536 points (256 x 256), coincided exactly with the values obtained directly from the equation for time-optimal control. This proves that the evolved program really implements the time-optimal control strategy for the problem.

The total evolution (100,000 cycles) time was approximately 3 hours. Around cycle number 7,000, which occurred within about 12 minutes, the first individuals that were able to center the car for all the training cases, that is, the individuals whose number of timeouts equaled zero, began to appear. From this point on, it was noticed that the evolution continues to reduce the

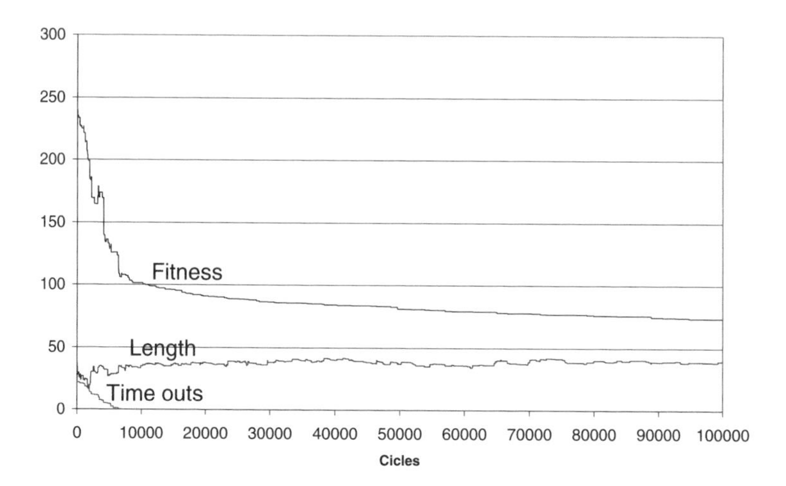

Fig. 9.10. Evolution of the cart-centering problem, with control over the length of the individuals

individuals' fitness, and the first individuals that represented the time-optimal control strategy appeared around cycle number 80,000.

For the purpose of evaluating the evolved program by means of the comparison mentioned above, its operation was simulated with the use of the Microchip MPLAB simulator program. To this end, the evolved code was inserted in the program shown in listing 9.1. The objective of this program is to create a main loop that continuously reads the input ports and updates the PIC's output port, its kernel being the evolved program, which has 25 instructions. It may be observed that the evolution did not require the use of auxiliary variables.

Listing 9.1. Listing of the program that contains the code evolved for the inverted pendulum problem

```
#INCLUDE <P18F452.INC>

  CBLOCK 0x08                ;Start of user memory
         X                   ;Position
         V                   ;Velocity
         U                   ;Control
  ENDC                       ;End of memory block

  ORG 0x00                   ;Reset vector
  GOTO BEGIN                 ;Initial execution address

BEGIN
  SETF       TRISB           ;TRISD:=11111111b
```

```
                              ;Define all port B pins
                              ;as inputs
    SETF        TRISC         ;TRISC:=11111111b
                              ;Define all port C pins
                              ;as inputs
    CLRF        TRISD         ;TRISB:=00000000b
                              ;Define all port D pins
                              ;as outputs

LOOP
    CLRF        W             ;W:=0
                              ;Initialize W register
    CLRF        PRODH         ;PRODH:=0
                              ;Initialize PRODH register
    CLRF        PRODL         ;PRODL:=0
                              ;Initialize PRODL register
    MOVFF       PORTB, X      ;X:=PORTD
                              ;Read port D (position)
    MOVFF       PORTC, V      ;V:=PORTC
                              ;Read port C (velocity)

    BTFSC       X,7           ;Start of evolved code
    SUBWF       V,W
    BTFSC       V,7
    SUBWF       V,F
    MULWF       X
    INCF        V,W
    ADDWF       X,F
    ADDWF       X,F
    MULWF       PRODH,W
    SUBWF       PRODH,W
    MULWF       X
    ADDWF       X,F
    ADDWF       X,F
    SUBWF       PRODH,W
    BTFSC       V,7
    ADDWF       X,F
    BTFSC       X,7
    SETF        U
    MULWF       PRODH
    SUBWF       PRODH,W
    ADDWF       X,F
    MULWF       X
    MULWF       PRODH
    BTFSC       PRODH,7
```

```
SETF      U           ;End of evolved code

MOVFF     U, PORTD    ;PORTD:=U
                      ;Write to port D (control)
GOTO      LOOP        ;Return to beginning of loop
END
```

9.7.2 Inverted Pendulum

The problem of balancing an inverted pendulum in a minimum amount of time by applying a bang-bang force in one or in another direction is quite a familiar time-optimal control problem and involves an inherently unstable mechanical system. The dynamics of this problem is nonlinear and the problem of balancing the pendulum has been approached in several ways. Some of the approaches involve linear controllers and were not very successful [17]. Recently, nonlinear controllers for this problem were studied by researchers in artificial intelligence, evolutionary computing [2][8][11][14] and neural networks [1]. These nonlinear controllers offer better possibilities for obtaining an optimal performance in terms of the time required for achieving the objective state and the amount of effort that is expended.

The inverted pendulum problem bears a certain similarity to the cart-centering problem in the sense that it involves a car with mass m_c moving along a one-dimensional frictionless track. Additionally, there is an inverted pendulum with mass m_p whose axis is on top of the car. The pendulum has an angle θ and an angular velocity ω. The distance from the mass center of the pendulum to its axis is λ.

There is a control variable for this system: a force F of fixed magnitude (bang-bang force) that can be applied to the mass center of the car at each time interval so as to accelerate it in the positive or negative directions along the track.

This system is shown in figure 9.11 and has four state variables: the car's position x along the axis, the car's velocity v, the angle θ of the pendulum (measured vertically) and the angular velocity ω of the pendulum. In the same manner as [14] and [8], this case study considers a particular version of this problem in which only three of the state variables are controlled and variable x relative to position is not included.

In this hypothetical experiment, the PIC takes the car's velocity, the pendulum's angle and angular velocity as its inputs and returns the control action according to the values of these inputs. For obtaining the value relative to the car's velocity $v(t)$, the signal originating from a potentiometer, which is attached to one of the wheels, passes through a differentiator that has been calibrated to provide a value between -0.2 and 0.2 volts to represent the value of $v(t)$ between -0.2 and 0.2 meters per second. This analog signal is then converted to digital and supplied to the PIC through port D. The value relative

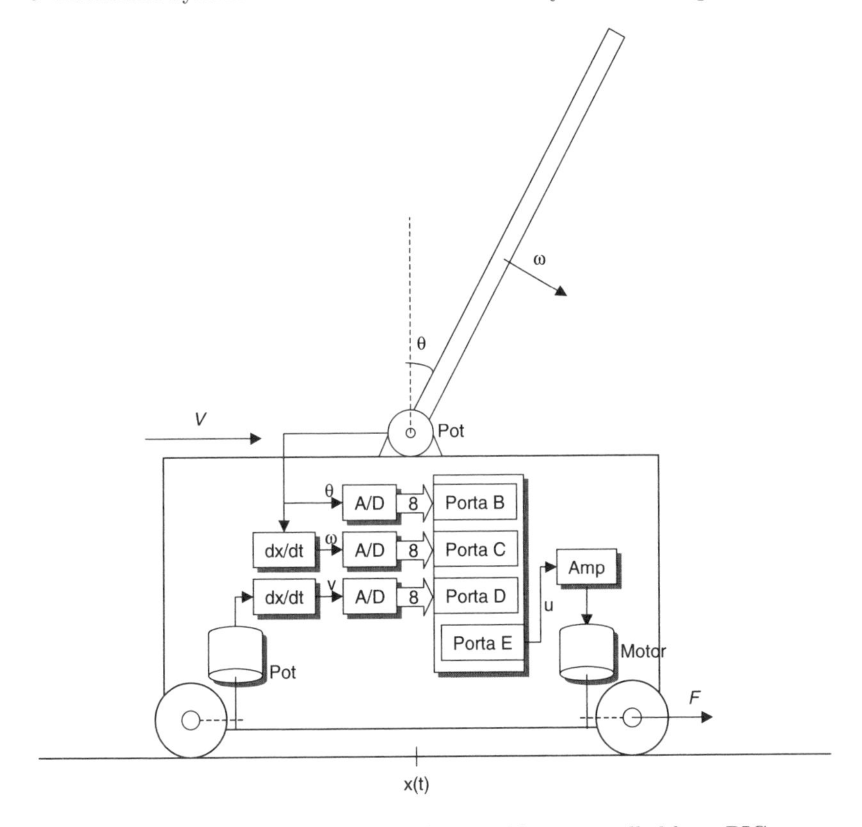

Fig. 9.11. The inverted pendulum problem controlled by a PIC

to the angle $\theta(t)$ is obtained by means of a potentiometer (attached to the pendulum's axis) that is fed and calibrated so as to provide a value between -0.2 and 0.2 volts to represent the value of $\theta(t)$ between -0.2 and 0.2 radians. This analog voltage value is then converted to digital and is supplied to the PIC through port B. For obtaining the value relative to the pendulum's angular velocity $\omega(t)$, the signal originating from the potentiometer passes through a differentiator that has been calibrated to provide a value between -0.2 and 0.2 volts to represent the value of $\omega(t)$ between -0.2 and 0.2 radians per second. This analog signal is then converted to digital and is supplied to the PIC through port C. The PIC performs the control $u(t)$ of the car through the pin relative to the least significant bit of port E. Its value is supplied to the current amplifier which, upon receiving the low logical value (0V), activates the motor so as to impart to the car a force in the negative direction, and upon receiving the high logical level (5V), activates it imparting a force in the positive direction to the car.

At each time interval, the selection of the control variable $u(t)$, which indicates in which direction the force F will be applied, at instant t, brings about a change in the system's state variables at instant $t+1$. This system's state transitions are expressed by nonlinear differential equations. At each discrete time interval t, the current state of the system and the force that is being applied at that interval determine the system's state at the following instant.

The pendulum's angular acceleration $\Phi(t)$ at instant t is given by [1] as

$$\Phi(t) = \frac{g\sin\theta + \cos\theta \frac{-F - m_p\lambda\omega\theta^2\sin\theta}{m_c+m_p}}{\lambda(\frac{4}{3} - \frac{m_p\cos^2\theta}{m_c+m_p})} \tag{9.6}$$

The constants are: the car mass ($m_c = 0.9$ kilogram), the pendulum mass ($m_p = 0.1$ kilogram), the gravity ($g = 1.0$ meter/sec2), the time interval ($t = 0.02$ second) and the length of the pendulum ($\lambda = 0.8106$ meter). The pendulum's angular velocity $w(t+1)$ at instant $t+1$ is therefore

$$\omega(t+1) = \omega(t) + \tau\Phi(t) \tag{9.7}$$

Thus, as a result of the angular acceleration $F(t)$, the angle $\theta(t+1)$ at instant $t+1$ is, with the use of Euler's integration approximation,

$$\theta(t+1) = \theta(t) + \tau\omega(t) \tag{9.8}$$

The acceleration $a(t)$ of the car on the track is given by

$$a(t) = \frac{F + m_p\lambda(\theta^2\sin\theta - \omega\cos\theta)}{m_c + m_p} \tag{9.9}$$

The velocity $v(t+1)$ of the car on the track at instant $t+1$ is therefore

$$v(t+1) = v(t) + \tau a(t) \tag{9.10}$$

Evaluation Function

The evaluation function of this case study is similar to the one in the previous study. The fitness of an individual is the total amount of time needed for balancing the pendulum, given some initial condition points (v, θ, ω) in the domain that has been specified for the problem. For the purpose of computing this total time, the individual must be submitted to a test in which it must be executed for all the training cases. The training cases are evenly spaced selected values of the state variables within the specified domain. In this case, $k = 27$ values were selected, 3 for $v(0)$, 3 for $\theta(0)$ and 3 for $\omega(0)$, and these three variables took values: -0.2, 0 and 0.2. In this manner, an individual's performance is tested for each one of these 27 training cases and the total amount of time is calculated.

The plant's simulation time is 6 seconds and if this period is exceeded, a timeout is caused. This simulation occurs at intervals of 0.02 second, that is, this loop is executed up to a maximum of 300 times. It is considered that the plant has reached its state of equilibrium when the square root of the sum of the squares of the car's velocity, of the angle and of the angular velocity of the pendulum is lower than 0.07.

Experiment

Table 9.4 summarizes this experiment and describes its main features and parameters.

Table 9.4. Summary of the evolutionary experiment with the inverted pendulum problem

Objective:	To synthesize an assembly language program that enables a PIC18F452 MC to control a car on a track so as to lead an inverted pendulum to a state of equilibrium.
Terminal set:	V (car's velocity), O (angle), w (angular velocity) and U (control); A1 and A2 (auxiliary); PRODH and PRODL (multiplier hardware registers); STATUS (flag register).
Function set:	The 22 PIC instructions described in Table 9.2.
Training cases:	27 initial condition points (v, θ, ω), selected and evenly spaced, between -0.2 and 0.2. There are 3 values for each variable: -0.2; 0 and 0.2.
Hits:	Number of training cases that did not cause a timeout.
Fitness:	Sum of the time, over the 27 training cases, taken to equilibrate the car. When a training case causes a timeout, the contribution is of 6 seconds.
Evaluation:	The least number of timeouts. As a first tie-breaking criterion, the lowest fitness. As a second tie-breaking criterion, the shortest length of the individual.
Main parameters:	M = 1000. Steady state. Number of processed individuals: 400,000, which is equivalent to G = 400. $p_c = 0.8$, $p_m = 0.4$.

The process adopted for evaluating each individual in this case study was the same as the one in the previous study. In other words, a significant improvement was observed in evolutionary performance when the individuals were evaluated with respect to the number of timeouts first, and in case there was a tie, with regard to the fitness itself. The heuristic on control over the length of the individuals was also used here since the experiments that were carried out in the previous case studies had demonstrated that it was satisfactory.

This problem represented increased complexity for the GP, since the system has three state variables to be controlled and is inherently unstable. As a result, it became necessary to increase the size of the population to M = 1,000 individuals. Each evolution was carried out up to cycle 100,000, hence, 400,000 individuals were evaluated. If the GP were generation-based, this would be equivalent to G = 400 generations of 1,000 individuals.

Figure 9.12 shows an evolutionary graph that represents the result of the average of 10 executions of the GP.

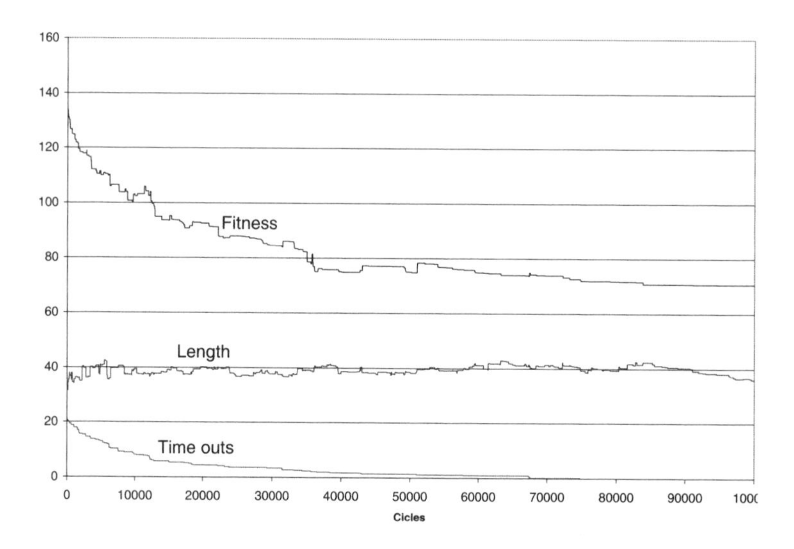

Fig. 9.12. Evolution of the inverted pendulum problem, with control over the length of the individuals.

There is no known solution to this problem, neither is there any specific test that might be applied with a view to establishing that an apparent solution that has been obtained is optimal. A pseudo-optimal strategy developed in [15] served as an approximate reference for determining the possible attainment of a suboptimal value for time. This pseudo-optimal strategy is an approximated solution to a linear simplification of the problem. When the initial conditions consist of the eight vertices of cube $v - \theta - \omega$, it takes this strategy an average of 2.96 seconds to balance the system. The evolved program, when submitted to the same points, balances the system in 2.84 seconds on the average. This shows that the GP was able to synthesize a solution that was better than the best solution obtained mathematically.

The best individual that was evolved in all the executions presented a fitness of approximately 67 seconds for the 27 training cases. The total evolution (100,000 cycles) time was approximately 5 hours. Around cycle number 66,000, which occurred within about 3 hours and 20 minutes, the first individuals that were able to balance the system for all the training cases, in other words, the individuals whose number of timeouts equaled 0, began to appear. From this point on, it was noticed that the evolution continues to reduce the individuals' fitness, and the first individuals that represented the evolved individual's control strategy appeared around cycle number 78,000.

For the purpose of validating the evolved program, its operation was simulated with the use of the Microchip MPLAB simulator program. To this end, the evolved code was inserted in the program shown in listing 9.2. The objective of this program is to create a main loop that continuously reads the input ports and updates the PIC's output port, its kernel being the evolved program, which has 25 instructions. It may be observed that the evolution did not require the use of auxiliary variables.

Listing 9.2. Listing of the program that contains the code evolved for the inverted pendulum problem

```
#INCLUDE <P18F452.INC>

   CBLOCK 0x08                    ;Start of user memory
            O                     ;Angle
            w                     ;Angular velocity
            V                     ;Car velocity
            U                     ;Control
   ENDC                           ;End of memory block

   ORG      0x00                  ;Reset vector
   GOTO     BEGIN                 ;Initial execution address

BEGIN
   SETF     TRISB                 ;TRISB:=11111111b
                                  ;Define all port B pins
                                  ;as inputs
   SETF     TRISC                 ;TRISC:=11111111b
                                  ;Define all port C pins
                                  ;as inputs
   SETF     TRISD                 ;TRISD:=11111111b
                                  ;Define all port D pins
                                  ;as inputs
   CLRF     TRISE                 ;TRISB:=00000000b
                                  ;Define all port E pins
                                  ;as outputs

LOOP
   CLRF     W                     ;W:=0
                                  ;Initialize W register
   CLRF     PRODH                 ;PRODH:=0
```

```
                                ;Initialize PRODH register
CLRF        PRODL               ;PRODL:=0
                                ;Initialize PRODL register
MOVFF       PORTB, O            ;O:=PORTB
                                ;Read port B (angle)
MOVFF       PORTC, w            ;w:=PORTC
                                ;Read port C (angular velocity)
MOVFF       PORTD, V            ;V:=PORTD
                                ;Read port D (car velocity)

ADDWFC      PRODH, F            ;Start of evolved code
BTFSC       STATUS, 0
RLCF        PRODL, F
RLCF        w, F
RLCF        O, W
ADDWF       w, F
ADDWFC      PRODH, F
ADDWF       w, W
ADDWFC      V, W
SUBWF       PRODL, F
MOVFF       w, W
RLCF        U, F
SUBWF       PRODH, F
RLCF        PRODL, F
ADDWF       V, W
ADDWFC      V, W
RLCF        O, W
ADDWFC      PRODL, F
CPFSGT      V
ADDWFC      PRODL, F
CPFSEQ      PRODL
CPFSGT      V
SUBWF       PRODH, W
BTFSS       U, 7
RLCF        U, F                ;End of evolved code

MOVFF       U, PORTE            ;PORTE:=U
                                ;Write to port E (control)
GOTO        LOOP                ;Return to beginning of loop
END
```

9.8 Summary

Systems based on genetic programming make it possible to automatically synthesize programs that implement optimal-time control strategies for microcontrollers directly in assembly language. Within this application class, the best model is the one used in linear genetic programming, in which each chromosome is represented by an instruction list. This model makes it possible to evolve programs in assembly language, which is an imperative language. This would not be possible with the use of a tree-based representation because such representations evolve programs in functional languages only.

The evolution of programs directly in assembly language has proved to be indispensable for two main reasons:

1. Because it represents a device's lowest possible programming level, it enabled evolution to find solutions with optimized codes since the evaluation functions excelled in programs with smaller amounts of instructions;
2. In terms of evolution time, it made the practical application of the system a viable alternative, since evolution in a higher-level language, such as C, for example, would require the compilation stage, which is executed at each evaluation, and this would represent a significant increase in the total evolution time and make it impossible to use the system in practice.

The case studies discussed here represented the problems one typically encounters when dealing with the control of physical systems, such as the fact that the models are nonlinear and the problem of controlling several state variables through actions performed on one control variable.

As for the final results, the experiments with the case studies demonstrated that the automatic synthesis system yields a good performance since it was able to automatically synthesize a control program whose performance was at least as good as the one found in the literature as a performance quality benchmark. The most important aspect of this approach is that the solutions found for the control problems were obtained already in the system's final implementation format, that is to say, a program for a microcontroller. This result was obtained directly from the mathematical models of the systems that were to be controlled, and other possible intermediate stages between the modeling and implementation stages became unnecessary. In other words, this methodology eliminates the need for the systems designer to find an optimal or suboptimal solution for the system mathematically and also eliminates the manual task of creating a program that implements this solution.

With regard to synthesis time, this methodology also proves that it can compete with a human designer because it is able to provide a solution in a few hours. It is important to point out that the development of control programs involving such complex systems as the ones in the case studies presented here may require a few days to be completed.

The total time spent in synthesis is still considered relatively long. There are several alternatives for reducing it.

First, develop an algorithm that identifies the parts of the program that do not present any influence on the calculation of the output(s), for all the possible inputs (introns). This algorithm would be based on the study by Brameier & Banzhaf [6], where the complexity level of the algorithm developed is, in the worst case, $O(n)$, n being the length of the individual. On the one hand, this approach would reduce the total evolutionary time because the introns would no longer be interpreted by the MC simulator. On the other, it would improve the throughput of the mutation operator, which would only be applied in the effective code sections and thus increase the velocity of the evolution.

Another possibility is to investigate the use of several demes [20] so as to delay the convergence of the GP as much as possible, since a conversion at a later time may represent the synthesis of programs with more optimized control performances. An additional alternative that may be considered is the use of MC instructions that make it possible to create execution loops that might contribute in some additional way to the quality of the evolved programs. This would also involve studying a manner by which to deal with undesirable infinite loops that might occur.

An additional alternative that may be considered is the use of MC instructions that make it possible to create execution loops that might contribute in some additional way to the quality of the evolved programs. This would also involve studying a manner by which to deal with undesirable infinite loops that might occur.

References

1. C. W. Anderson. Learning to control an inverted pendulum using neural networks. *IEEE Control Systems Magazine*, pages 31–37, 1989.
2. Peter J. Angeline. An alternative to indexed memory for evolving programs with explicit state representations. In John R. Koza, Kalyanmoy Deb, Marco Dorigo, David B. Fogel, Max Garzon, Hitoshi Iba, and Rick L. Riolo, editors, *Genetic Programming 1997: Proceedings of the Second Annual Conference*, pages 423–430, Stanford University, CA, USA, 1997. Morgan Kaufmann.
3. K. J. Astrm and B. Wittenmark. *Computer-Controlled systems: theory and design*. Prentice-Hall, New Jersey, 1997.
4. M. Athans and P. L. Falb. *Optimal Control: An Introduction to the Theory and Its Applications*. McGraw-Hill Book Company, New York, 1966.
5. Wolfgang Banzhaf, Peter Nordin, Robert E. Keller, and Frank D. Francone. *Genetic Programming – An Introduction; On the Automatic Evolution of Computer Programs and its Applications*. Morgan Kaufmann, dpunkt.verlag, 1998.
6. Markus Brameier and Wolfgang Banzhaf. Effective linear genetic programming. Technical report, Department of Computer Science, University of Dortmund, 44221 Dortmund, Germany, 2001.
7. Markus Brameier, Wolfgang Kantschik, Peter Dittrich, and Wolfgang Banzhaf. SYSGP – A C++ library of different GP variants. Technical Report CI-98/48, Collaborative Research Center 531, University of Dortmund, Germany, 1998.

8. Kumar Chellapilla. Automatic generation of nonlinear optimal control laws for broom balancing using evolutionary programming. In *Proceedings of the 1998 IEEE World Congress on Computational Intelligence*, pages 195–200, Anchorage, Alaska, USA, 1998. IEEE Press.
9. Nichael Lynn Cramer. A representation for the adaptive generation of simple sequential programs. In John J. Grefenstette, editor, *Proceedings of an International Conference on Genetic Algorithms and the Applications*, pages 183–187, Carnegie-Mellon University, Pittsburgh, PA, USA, 24-26 July 1985.
10. Ronald L. Crepeau. Genetic evolution of machine language software. In Justinian P. Rosca, editor, *Proceedings of the Workshop on Genetic Programming: From Theory to Real-World Applications*, pages 121–134, Tahoe City, California, USA, 1995.
11. D. B. Fogel. A 'correction' to some cart-pole experiments. In T. Baeck L. J. Fogel, P. J. Fogel, editor, *Evolutionary Programming V: Proceedings of the Fifth Annual Conference on Evolutionary Programming*, pages 67–71, Cambridge, MA, 1996. MIT Press.
12. Lorenz Huelsbergen. Toward simulated evolution of machine language iteration. In John R. Koza, David E. Goldberg, David B. Fogel, and Rick L. Riolo, editors, *Genetic Programming 1996: Proceedings of the First Annual Conference*, pages 315–320, Stanford University, CA, USA, 28–31 July 1996. MIT Press.
13. Kenneth E. Kinnear, Jr. Evolving a sort: Lessons in genetic programming. In *Proceedings of the 1993 International Conference on Neural Networks*, volume 2, pages 881–888, San Francisco, USA, 1993. IEEE Press.
14. John R. Koza. *Genetic Programming: On the Programming of Computers by Means of Natural Selection.* MIT Press, Cambridge, MA, USA, 1992.
15. John R. Koza and Martin A. Keane. Cart centering and broom balancing by genetically breeding populations of control strategy programs. In *Proceedings of International Joint Conference on Neural Networks*, volume I, pages 198–201, Washington, 1990. Lawrence Erlbaum.
16. Microchip. *PIC18FXX Data Sheet*, 2002. http://www.microchip.com.
17. S. Mori, H. Nishihara, and K. Furuta. Control of unstable mechanical system-control of pendulum. *Int. J. Control*, pages 673–692, 1976.
18. Peter Nordin. A compiling genetic programming system that directly manipulates the machine code. In Kenneth E. Kinnear, Jr., editor, *Advances in Genetic Programming*, chapter 14, pages 311–331. MIT Press, 1994.
19. Peter Nordin. *Evolutionary Program Induction of Binary Machine Code and its Application.* Krehl-Verlag, Mnster, Germany, 1997.
20. Peter Nordin, Frank Hoffmann, Frank D. Francone, Markus Brameier, and Wolfgang Banzhaf. AIM-GP and parallelism. In Peter J. Angeline, Zbyszek Michalewicz, Marc Schoenauer, Xin Yao, and Ali Zalzala, editors, *Proceedings of the Congress on Evolutionary Computation*, volume 2, pages 1059–1066, Mayflower Hotel, Washington D.C., USA, 1999. IEEE Press.
21. Katsuhiko Ogata. *Modern Control Engineering.* Prentice-Hall, 1997.

Index

Index

Reviewer List